Instructor's Resource Manual and Test Bank
to Accompany Ahrens'
Meteorology Today

Seventh Edition

Jonathan D. W. Kahl
University of Wisconsin – Milwaukee

THOMSON

BROOKS/COLE

Australia • Canada • Mexico • Singapore • Spain • United Kingdom • United States

Printed in the United States of America

1 2 3 4 5 6 7 05 04 03 02 01

Printer: Victor Graphics

0-534-39776-X

For more information about our products, contact us at:
Thomson Learning Academic Resource Center
1-800-423-0563

For permission to use material from this text, contact us by:
Phone: 1-800-730-2214
Fax: 1-800-731-2215
Web: http://www.thomsonrights.com

Asia
Thomson Learning
5 Shenton Way #01-01
UIC Building
Singapore 068808

Australia
Nelson Thomson Learning
102 Dodds Street
South Street
South Melbourne, Victoria 3205
Australia

Canada
Nelson Thomson Learning
1120 Birchmount Road
Toronto, Ontario M1K 5G4
Canada

Europe/Middle East/South Africa
Thomson Learning
High Holborn House
50/51 Bedford Row
London WC1R 4LR
United Kingdom

Latin America
Thomson Learning
Seneca, 53
Colonia Polanco
11560 Mexico D.F.
Mexico

Spain
Paraninfo Thomson Learning
Calle/Magallanes, 25
28015 Madrid, Spain

Contents

Chapter	Title	Page
	Preface	i
Chapter 1	The Earth and its Atmosphere	1
Chapter 2	Energy: Warming the Earth and the Atmosphere	22
Chapter 3	Seasonal and Daily Temperatures	46
Chapter 4	Light, Color, and Atmospheric Optics	69
Chapter 5	Atmospheric Moisture	88
Chapter 6	Condensation: Dew, Fog, and Clouds	104
Chapter 7	Stability and Cloud Development	129
Chapter 8	Precipitation	146
Chapter 9	The Atmosphere in Motion: Air Pressure, Forces, and Winds	161
Chapter 10	Wind: Small-Scale and Local Systems	185
Chapter 11	Wind: Global Systems	205
Chapter 12	Air Masses and Fronts	220
Chapter 13	Middle Latitude Cyclones	240
Chapter 14	Weather Forecasting	252
Chapter 15	Thunderstorms and Tornadoes	264
Chapter 16	Hurricanes	288
Chapter 17	Air Pollution	301
Chapter 18	Global Climate	316
Chapter 19	Climate Change	332
Appendix:	WebTutor on WebCT and Blackboard Teaching Tips	

Preface

This manual provides supplementary information to assist instructors using Ahrens' *Meteorology Today*, 7th Edition. Each chapter of this document contains the following sections:

- **Summary** - A quick reference to the principle topics addressed in the chapter. This section may assist instructors who choose not to use all chapters of the text.

- **Key Terms** - A listing of key terminology used in the chapter.

- **Teaching Suggestions** - Assorted demonstrations, discussion topics, and miscellaneous ideas for classroom presentation. Suggestions linked to activities on the BlueSkies cdrom are indicated by the BlueSkies icon.

- **Student Projects** - This section describes for hands-on projects to supplement lecture and reading material. Projects linked to activities on the BlueSkies cdrom are indicated by the BlueSkies icon.

- **Answers to Questions for Thought** - Questions for Thought occur at the end of each chapter in the Ahrens *Meteorology Today*, 7th Edition text.

- **Answers to Problems and Exercises** - Problems and Exercises occur at the end of each chapter in the Ahrens *Meteorology Today*, 7th Edition text.

- **Multiple Choice Exam Questions** - Between 40 and 95 multiple choice questions, with answers, are provided for each chapter.

- **Essay Exam Questions** - Approximately 10-15 essay questions are provided for each chapter.

What's New in the 7th Edition

All chapters of *Meteorology Today*, 7th Edition have been updated and revised. Major additions include:

- new material on greenhouse gases, wind chill, National Ambient Air Quality Standards, the Pacific-Decadal Oscillation, and upper-level fronts

- recent research results on climate change

- major reorganization of "Weather Forecasting" and "Thunderstorms and Tornadoes" chapters

- new Focus Section topics, including Drylines, Martian Dust Storms and the Southwest U.S. Monsoon

- updated descriptions of meteorological obserrving systems, including ASOS

- updated color satellite imagery

Chapter 1
The Earth and its Atmosphere

Summary

This introductory chapter presents a broad overview of the physical structure of the atmosphere and its weather. The chapter begins with a discussion of the present composition of the earth's atmosphere. The important and varied roles played by water vapor, which is a source of precipitation and latent heat energy as well as being the most important greenhouse gas, are given particular attention. Current concern over increasing concentrations of another constituent, carbon dioxide, and its possible effect on global climate are also examined. The student will see that the observed increase in CO_2 is a result of an imbalance between processes of release and removal. The principle atmospheric pollutants, including ozone, are listed but are covered in greater detail in a later chapter.

The concepts of air density and air pressure are introduced and their variation with altitude is examined. A vertical profile of temperature shows that the atmosphere can be divided into several layers with distinct properties. The ionosphere is described briefly, as are the atmospheric characteristics of the other planets.

Finally, the student is introduced to the elements that constitute weather and will see how weather conditions might be depicted on a surface weather map and in a photograph from a geostationary satellite. The chapter includes discussions of the history of and careers in meteorology, and ends with a description of the many ways that weather and climate can affect our lives and interests.

Key Terms

radiant energy
atmosphere
atoms
molecules
ions
proton
neutron
electron
nitrogen
oxygen
water vapor
condensation
evaporation
latent heat
greenhouse gas
carbon dioxide
deforestation
photosynthesis
global warming
methane (CH_4)
nitrous oxide (N_2O)
trace gases
ozone
photochemical smog
aerosols
pollutants
nitrogen dioxide (NO_2)
carbon monoxide (CO)
hydrocarbons
sulfur dioxide (SO_2)
acid rain
outgassing
photodissociation
gravity
density

pressure
atmospheric pressure
air pressure
millibar
bar
newton
pascal
lapse rate
average lapse rate
troposphere
radiosonde
isothermal
stratosphere
tropopause
jet streams
rawinsonde
sounding
temperature inversion
sudden warming
mesosphere
stratopause
hypoxia
thermosphere
mesopause
mean free path
exosphere
homosphere
heterosphere
ionosphere
D region
oxides of nitrogen
weather
weather elements
climate

meteorology
Meteorologica
Tiros I
geostationary satellite
meridians
longitude
latitude
middle latitudes
middle latitude
  cyclonic storm
extratropical cyclone
hurricane
hurricane eye
thunderstorm
tornado
low pressure (center)
high pressure(center)
anticyclones
wind
wind direction
front
cold front
warm front
occluded front
wind chill
frost bite
hypothermia
heat exhaustion
heat stroke
severe thunderstorms
flash floods
downburst
wind shear
NOAA weather radio

Teaching Suggestions

1.   Some of the atmospheric pressure demonstrations described in Chapter 9 could be performed here also.

2.   Fill a wine glass completely with water and cover it with a piece of plastic (such as the lid from a tub of margarine), being careful to remove any air. Invert the glass. The water remains in the glass because the upward force on the cover due to the pressure of the air is much stronger than the downward gravitational force on the water. The demonstration can be made much more convincing if a 4000 mL erlenmeyer flask is used instead of the wine glass. When full of water, the flask weighs approximately 10 pounds.

3.   Place a candle in the center of a dish and partly fill the dish with water. Light the candle and then cover it with a large jar or beaker. The flame will consume the oxygen inside the jar and reduce the pressure. Water will slowly flow into the jar to re-establish pressure balance. The change in volume will be close to 20%, the volume originally occupied by the oxygen in the air. This demonstration can be used to illustrate the concept of partial pressure, which is later used in the chapter on humidity. The students should also be asked what they think the products of the combustion might be and why these gases do not replace the oxygen and maintain the original pressure in jar. One of the combustion products is water vapor, which condenses as the air in the jar cools. Another combustion product is carbon dioxide, which presumably goes into solution.

4.   Students often confuse water vapor with liquid water. Students should understand that water vapor is an invisible gas. Haze, fog, clouds, and the steam from a boiling pot all become visible when water vapor condenses and forms small drops of <u>liquid</u> water. This can be easily demonstrated using a tea kettle, or by showing a video of water boiling in a tea kettle.

5.   The introductory explanations of the air motions associated with high and low pressure centers and fronts, make this a good place to begin to show and discuss satellite photographs, loops, and surface weather maps. Many students have occasion to watch television weather broadcasts. Being able to observe and understand weather phenomena on their own may heighten interest in the subject. Download a few satellite loops off the Internet and discuss the air motions.

 Blue Skies 6.   Use the Atmospheric Basics/Layers of the Atmosphere section of the BlueSkies cdrom to demonstrate that most of the atmosphere's water vapor is in the lower half of the troposphere.

Student Projects

 Blue Skies 1.   Have the students mark the positions of fronts and pressure systems for each day on an outline map of the United States. (This information can be obtained from the daily newspaper, the TV news broadcast, or from the Weather Analysis/Find the Front section of the BlueSkies cdrom.) Have students do this for a week at a time, noting the general movement of these systems.

2.   Have students compose a one-week journal, including daily newspaper weather maps and weather forecasts. Have the students provide a commentary for each day as to the coincidence of actual and predicted weather.

3.   Have students keep a daily record of weather observations, especially significant changes in the weather. Then, periodically, the instructor can supply mean daily data such as high and low temperatures, pressure, dew point, wind speed, cloud cover and precipitation amounts. The students should plot this data and annotate the graph with their observations. Students can use their graphs to experimentally test concepts developed in class. After studying Chapter 1, for example, students might try to determine whether periods of stormy weather really are associated with lower-than-average surface pressure.

Blue Skies 4.   Use the Atmospheric Basics/Layers of the Atmosphere section of the BlueSkies cdrom to identify the altitude of the tropopause at five different cities.

Answers to Questions for Thought

1.   Weather: (b), (d), (g). Climate: (a), (c), (e), (f), and (h).

2.   Because weather is largely advective, meaning that weather conditions often move from place to place. It makes sense to call a wind "westerly" if it is blowing from the west, thus bringing weather conditions from the west.

3.   (a) 0.5 ATM and 0.1 ATM are equal to about 500 mb and 100 mb, respectively. From the figure, 500 mb is located at an altitude of about 5.5 km (3.5 miles); 100 mb is found at an altitude of about 16 km (10 miles). The surface pressure on Mars, 0.007 ATM, is about 7 mb. A pressure of 7 mb would be found near 35 km (22 miles) altitude in the earth's atmosphere.

4.   Your stomach would expand, because the pressure outside your body would be several orders of magnitude less than the pressure inside your body.

Multiple Choice Exam Questions

1.   The primary source of energy for the earth's atmosphere is:
       a. energy from within the earth
       b. the sun
       c. erupting volcanoes
       d. lightning discharges associated with thunderstorms
       e. latent heat released during the formation of hurricanes

ANSWER: b

2. The most abundant gases in the earth's atmosphere by volume are:
    a. carbon dioxide and nitrogen
    b. oxygen and water vapor
    c. nitrogen and oxygen
    d. oxygen and helium
    e. oxygen and ozone

ANSWER: c

3. A single breath of air contains about
    a. 10^2 molecules
    b. 10^2 ions
    c. 10^{22} molecules
    d. 10^{22} ions

ANSWER: c

4. Water vapor is:
    a. a gas
    b. a cloud droplet
    c. a rain drop
    d. a snowflake

ANSWER: a

5. In a volume of air near the earth's surface, ___ occupies 78 percent
    and ___ nearly 21 percent.
    a. nitrogen, oxygen
    b. hydrogen, oxygen
    c. oxygen, hydrogen
    d. nitrogen, water vapor
    e. hydrogen, helium

ANSWER: a

6. The negatively-charged particle that circles around the nucleus of an atom is:
    a. an electron
    b. a neutron
    c. an aerosol
    d. a proton

ANSWER: a

7.   Which of the following is considered a variable gas in the earth's atmosphere?
      a. water vapor
      b. nitrogen
      c. oxygen
      d. argon

ANSWER: a

8.   The gas that shows the most variation from place to place and from time to time in the lower atmosphere:
      a. ozone (O_3)
      b. carbon dioxide (CO_2)
      c. water vapor (H_2O)
      d. methane (CH_4)
      e. argon (Ar)

ANSWER: c

9.   Water vapor:
      a. is invisible
      b. colors the sky blue
      c. makes clouds white
      d. is very small drops of liquid water

ANSWER: a

10.   Typically, water vapor occupies about what percentage of the air's volume near the earth's surface?
      a. about 78%
      b. about 21%
      c. close to 10%
      d. less than 4%

ANSWER: d

11.   The only substance near the earth's surface that is found naturally in the atmosphere as a solid, liquid, and a gas:
      a. carbon dioxide
      b. water
      c. molecular oxygen
      d. ozone
      e. carbon

ANSWER: b

12. In the atmosphere, tiny solid or liquid suspended particles of various composition are called:
    a. aerosols
    b. carcinogens
    c. greenhouse gases
    d. microbes

ANSWER: a

13. The most abundant greenhouse gas in the earth's atmosphere:
    a. carbon dioxide (CO_2)
    b. nitrous oxide (N_2O)
    c. water vapor (H_2O)
    d. methane (CH_4)
    e. chlorofluorocarbons (CFCs)

ANSWER: c

14. Since the turn of this century, CO_2 in the atmosphere has:
    a. been increasing in concentration
    b. been decreasing in concentration
    c. remained at about the same concentration from year to year
    d. disappeared entirely

ANSWER: a

15. The greenhouse gas that has been increasing in concentration, at least partly due to deforestation, is:
    a. carbon dioxide (CO_2)
    b. chlorofluorocarbons (CFCs)
    c. water vapor (H_2O)
    d. ozone (O_3)

ANSWER: a

16. Which below is not considered a greenhouse gas?
    a. carbon dioxide (CO_2)
    b. nitrous oxide (N_2O)
    c. water vapor (H_2O)
    d. methane (CH_4)
    e. oxygen (O_2)

ANSWER: e

17. Which of the following processes acts to remove carbon dioxide from the atmosphere?
    a. lightning
    b. deforestation
    c. photosynthesis
    d. burning fossil fuels

ANSWER: c

18. The outpouring of gases from the earth's hot interior is called:
    a. evaporation
    b. outgassing
    c. photodissociation
    d. the hydrologic cycle

ANSWER: b

19. The earth's first atmosphere was composed primarily of:
    a. carbon dioxide and water vapor
    b. hydrogen and helium
    c. oxygen and water vapor
    d. argon and nitrogen

ANSWER: b

20. The primary source of oxygen for the earth's atmosphere during the past half billion years or so appears to be:
    a. volcanic eruptions
    b. photosynthesis
    c. photodissociation
    d. exhalations of animal life
    e. transpiration

ANSWER: b

21. The most abundant gas emitted from volcanoes is:
    a. nitrogen
    b. sulfur dioxide
    c. helium
    d. carbon dioxide
    e. water vapor

ANSWER: e

22. Hypoxia is a condition caused by:
    a. lack of oxygen going to the brain
    b. over exposure to ultraviolet radiation
    c. the combined effects of heat and humidity
    d. rapid changes in atmospheric pressure
    e. extreme cold

ANSWER: a

23. This holds a planet's atmosphere close to its surface:
    a. radiation
    b. gravity
    c. cloud cover
    d. moisture
    e. pressure

ANSWER: b

24. The amount of force exerted over an area of surface is called:
    a. density
    b. weight
    c. temperature
    d. pressure

ANSWER: d

25. Much of Tibet lies at altitudes over 18,000 feet where the pressure is about 500 mb. At such altitudes, the Tibetans are above roughly:
    a. 10% of the air molecules in the atmosphere
    b. 25% of the air molecules in the atmosphere
    c. 50% of the air molecules in the atmosphere
    d. 75% of the air molecules in the atmosphere

ANSWER: c

26. Which of the following are not units of pressure?
    a. millibars
    b. newtons
    c. inches of mercury (Hg)
    d. pascals

ANSWER: b

27. The unit of pressure most commonly found on a surface weather map:
    a. inches of mercury (Hg)
    b. millibars
    c. pounds per square inch
    d. millimeters of mercury (Hg)

ANSWER: b

28. Which of the following weather elements always decreases as we climb upward in the atmosphere?
    a. wind
    b. temperature
    c. pressure
    d. moisture
    e. all of the above

ANSWER: c

29. The number or mass of air molecules in a given space or volume is called:
    a. density
    b. pressure
    c. temperature
    d. weight

ANSWER: a

30. A planet whose atmosphere is mainly nitrogen and oxygen:
    a. Venus
    b. Mars
    c. Earth
    d. Jupiter
    e. Mercury

ANSWER: c

31. The atmosphere of __ is composed primarily of carbon dioxide (CO_2).
    a. Earth
    b. Mars
    c. Jupiter
    d. none of the above

ANSWER: b

32.   The gas responsible for the greenhouse effect on Venus:
        a. carbon dioxide (CO_2)
        b. oxygen (O_2)
        c. ozone (O_3)
        d. nitrogen (N_2)
        e. water vapor (H_2O)

ANSWER: a

33.   The planet with a strong greenhouse effect, whose surface temperature averages 480 °C (900 °F):
        a. Earth
        b. Venus
        c. Mars
        d. Pluto

ANSWER: b

34.   In the stratosphere, the air temperature normally:
        a. decreases with increasing height
        b. increases with increasing height
        c. both increases and decreases depending on the season
        d. cannot be measured

ANSWER: b

35.   The earth's atmosphere is divided into layers based on the vertical profile of:
        a. air pressure
        b. air temperature
        c. air density
        d. wind speed

ANSWER: b

36.   Carbon dioxide is a naturally-occurring component of the atmosphere.
        a. true
        b. false

ANSWER: a

37.   Almost all of the earth's weather occurs in the:
        a. exosphere
        b. stratosphere
        c. mesosphere
        d. thermosphere
        e. troposphere

ANSWER: e

38.   The most abundant gas in the <u>stratosphere</u> is:
        a. oxygen (O_2)
        b. nitrogen (N_2)
        c. carbon dioxide (CO_2)
        d. ozone (O_3)
        e. chlorofluorocarbons (CFCs)

ANSWER: b

39.   The hottest atmospheric layer is the:
        a. stratosphere
        b. mesosphere
        c. thermosphere
        d. troposphere

ANSWER: c

40.   Scientists are able to determine the air temperature in the thermosphere by:
        a. using radiosondes
        b. using temperature probes in orbiting satellites
        c. observing changes in satellite orbits
        d. direct measurements in manned, high-altitude balloons

ANSWER: c

41.   The atmospheric layer in which we live is called the:
        a. troposphere
        b. stratosphere
        c. thermosphere
        d. ionosphere
        e. exosphere

ANSWER: a

42.   The temperature of the tropopause:
        a. is close to the temperature at the earth's surface
        b. is much colder than the temperature at the earth's surface
        c. has never been measured
        d. is much warmer than the temperature at the earth's surface
        e. is nearly the same as the sun's temperature

ANSWER: b

43.   The instrument that measures temperature, pressure, and humidity at various altitudes in the atmosphere:
       a. barograph
       b. radiosonde
       c. aneroid barometer
       d. altimeter

ANSWER: b

44.   Warming in the stratosphere is mainly caused by:
       a. absorption of ultraviolet radiation by ozone
       b. release of latent heat energy during condensation
       c. chemical reactions between ozone and chlorofluorocarbons
       d. frictional heating caused by meteorites

ANSWER: a

45.   In a temperature inversion:
       a. air temperature increases with increasing height
       b. air temperature decreases with increasing height
       c. air temperature remains constant with increasing height
       d. it is warmer at night than during the day

ANSWER: a

46.   The rate at which temperature decreases with increasing altitude is known as the:
       a. temperature slope
       b. lapse rate
       c. sounding
       d. thermocline

ANSWER: b

47.   Atmospheric concentrations of carbon dioxide tend to go up and down throughout the course of a year.
       a. true
       b. false

ANSWER: a

48.   The main reason nighttime radio broadcasts can be sent over long distances is because:
       a. the low D-level region of the ionosphere is weaker at night
       b. there is less interference because many radio stations do not broadcast at night
       c. radio stations put out more power
       d. radio waves propagate more efficiently through cooler, high density air

ANSWER: a

49.   The electrified region of the upper atmosphere is called the:
      a. thermosphere
      b. mesosphere
      c. stratosphere
      d. ionosphere
      e. troposphere

ANSWER: d

50.   The ionosphere is an atmospheric layer that contains a high concentration of ions. An ion is:
      a. another term for ozone
      b. an atom or molecule that has lost or gained an electron
      c. atomic oxygen
      d. a radioactive element

ANSWER: b

51.   Most of the ionosphere is found in what atmospheric layer?
      a. troposphere
      b. stratosphere
      c. mesosphere
      d. thermosphere

ANSWER: d

52.   The gas that absorbs most of the harmful ultraviolet radiation in the stratosphere:
      a. water vapor
      b. nitrous oxide
      c. carbon dioxide
      d. ozone
      e. chlorofluorocarbons

ANSWER: d

53.   Which of the following equations is correct?
      a. Weight = mass ÷ gravity
      b. Weight = mass + gravity
      c. Weight = mass x gravity
      d. Weight = mass - gravity

ANSWER: c

54.	Meteorology did not become a genuine science until:
	a. Aristotle wrote <u>Meteorologica</u>
	b. the invention of weather instruments
	c. scientists discovered weather fronts
	d. computers were invented
	e. satellite data became available to the weather forecaster

ANSWER: b

55.	Which latitude belt best describes the middle latitudes?
	a. $20°$ to $80°$
	b. $10°$ to $35°$
	c. $20°$ to $35°$
	d. $40°$ to $70°$
	e. $30°$ to $50°$

ANSWER: e

56.	As altitude increases in the atmosphere, air density decreases _____ the decrease in air pressure.
	a. in a completely different way than
	b. much less than
	c. much more than
	d. in much the same way as

ANSWER: d

57.	The word "weather" is defined as:
	a. the average of the weather elements
	b. the climate of a region
	c. the condition of the atmosphere at a particular time and place
	d. any type of falling precipitation

ANSWER: c

58.	The wind direction is:
	a. the direction from which the wind is blowing
	b. the direction to which the wind is blowing
	c. always directly from high toward low pressure
	d. always directly from low toward high pressure

ANSWER: a

59. Meteorology is the study of:
    a. landforms
    b. the oceans
    c. the atmosphere
    d. outer space
    e. extraterrestrial meteoroids that enter the atmosphere

ANSWER: c

60. A south wind:
    a. blows from the north
    b. is any warm wind
    c. blows from the south
    d. is any moist wind

ANSWER: c

61. Storms vary in size (diameter). Which list below arranges storms from largest to smallest?
    a. hurricane, tornado, middle latitude cyclone, thunderstorm
    b. tornado, middle latitude cyclone, hurricane, thunderstorm
    c. hurricane, middle latitude cyclone, thunderstorm, tornado
    d. middle latitude cyclone, tornado, hurricane, thunderstorm
    e. middle latitude cyclone, hurricane, thunderstorm, tornado

ANSWER: e

62. A tropical storm system whose winds are in excess of 64 knots (74 mi/hr):
    a. thunderstorm
    b. anticyclone
    c. tornado
    d. extratropical cyclone
    e. hurricane

ANSWER: e

63. Middle latitude storms are also known as:
    a. anticyclones
    b. hurricanes
    c. extratropical cyclones
    d. tornadoes

ANSWER: c

64. A towering cloud, or cluster of clouds, accompanied by thunder, lightning, and strong, gusty winds:
    a. hurricane
    b. trough
    c. thunderstorm
    d. tornado

ANSWER: c

65. At night, when the weather is extremely cold and dry,
    a. atmospheric pressure increases with increasing altitude
    b. atmospheric pressure remains constant with increasing altitude
    c. atmospheric pressure decreases with increasing altitude
    d. atmospheric pressure first increases, then decreases with increasing altitude

ANSWER: c

66. In the middle latitudes of the Northern Hemisphere, surface winds tend to blow __ and __ around an area of surface low pressure.
    a. clockwise, inward
    b. clockwise, outward
    c. counterclockwise, inward
    d. counterclockwise, outward

ANSWER: c

67. In the middle latitudes of the Northern Hemisphere, surface winds tend to blow __ and __ around an area of surface high pressure.
    a. clockwise, inward
    b. clockwise, outward
    c. counterclockwise, inward
    d. counterclockwise, outward

ANSWER: b

68. Where cold surface air is replacing warm air, the boundary separating the different bodies of air is termed a:
    a. parallel of latitude
    b. tornado
    c. cold front
    d. warm front

ANSWER: c

69. The difference in altitude (i.e. the thickness) is greatest in the layer bounded by:
    a. 1 mb and 10 mb
    b. 101 mb and 110 mb
    c. 1001 mb and 1010 mb
    d. impossible to determine

ANSWER: a

70. On a weather map, sharp changes in temperature, humidity and wind direction are marked by:
    a. a front
    b. an anticyclone
    c. a ridge
    d. blowing dust

ANSWER: a

71. Which of the following is most likely associated with fair weather?
    a. high pressure area
    b. low pressure area
    c. a cold front
    d. a warm front

ANSWER: a

72. Areas of high atmospheric pressure are also known as:
    a. hurricanes
    b. middle latitude cyclonic storms
    c. troughs
    d. tornadoes
    e. anticyclones

ANSWER: e

73. Condensation is more likely to occur:
    a. when the air cools
    b. when the wind is calm
    c. when winds blow from the ocean over land
    d. at night

ANSWER: a

74. Clouds often form in the:
    a. rising air in the center of a low pressure area
    b. rising air in the center of a high pressure area
    c. sinking air in the center of a low pressure area
    d. sinking air in the center of a high pressure area

ANSWER: a

75. Generally, weather in the middle latitudes tends to move from ___ to ___.
    a. west to east
    b. east to west
    c. north to south
    d. south to north

ANSWER: a

76. Which relates to weather rather than climate?
    a. The average temperature for the month of January is 28 °F
    b. the lowest temperature ever recorded in Frozenlake,
       Minnesota is -57 °F
    c. the foggiest month of the year is December
    d. I like the warm, humid summers
    e. Outside it is cloudy and snowing

ANSWER: e

77. In an average year, more people die from ___ than from any other natural disaster.
    a. lightning
    b. earthquakes
    c. tornadoes
    d. flash floods and flooding
    e. droughts

ANSWER: d

78. At the 500 mb level, the amount of oxygen inhaled in a single breath is _____ of
    that inhaled at sea-level.
    a. about the same
    b. about one-quarter
    c. about one-half
    d. about three-quarters

ANSWER: c

79. Jupiter's "Great Red Spot" is
    a. a huge crater
    b. a huge spinning eddy
    c. a huge volcano
    d. a huge cloud of water vapor

ANSWER: b

80. Winds and temperatures throughout the troposphere and stratosphere are routinely measured by
    a. mercury barometer
    b. stethoscope
    c. ceilometer
    d. radiosonde

ANSWER: d

Essay Exam Questions

1. Describe the various types of storms found in the earth's atmosphere. Can you find any correlation between storm size and storm duration? What factors might determine a storm's severity?

2. What instruments are used in meteorology? What role did the discovery of instruments play in the emergence of the science of meteorology?

3. Briefly describe some of the historical events that helped meteorology progress as a natural science from Aristotle to the present day.

4. Under what circumstances might a person breathe stratospheric air? How often is it likely to happen in a student's lifetime.

5. What causes air pressure? Why does air pressure decrease with increasing altitude?

6. Describe some of the processes that release and remove carbon dioxide from the atmosphere. Is there any evidence that suggests that these processes are not in balance?

7. There is currently concern that the amount of ozone in the stratosphere may be decreasing. Why would a decrease in ozone concentration be important? Describe some of the effects that a decrease in ozone concentration might have.

8. If the air temperature at the surface (0 feet) is 60 °F, what would be the approximate air temperature at an altitude of 10,000 feet assuming an average atmospheric lapse rate of 3.6 °F per 1000 feet?

9. Draw a diagram showing how air temperature normally changes with height. Begin at the ground and end in the upper thermosphere. Be sure to label the four main layers. Give one important characteristic of each layer. Where on your diagram would the top of Mt. Everest, the ozone layer, and the ionosphere be found?

10. What are the principal gaseous components of the earth's atmosphere? Where do scientists believe these gases came from?

11. Why is there very little water vapor above the tropopause?

12.   What information might you find on a surface weather map that is not readily apparent on a satellite photograph? What information could a satellite photograph provide that a surface chart could not?

13.   Explain briefly why it is possible to transmit AM radio waves over larger distances at night than during the day.

Chapter 2
Energy: Warming the Earth
and the Atmosphere

Summary

This chapter begins with a definition of temperature and a comparison of the absolute, Celsius, and Fahrenheit temperature scales. Heat, the flow of energy between objects having different temperatures, occurs in the atmosphere by the processes of conduction, convection, and radiation. Air is a relatively poor conductor of heat but can transport heat efficiently over large distances by the process of convection. The latent heat energy associated with changes of phase of water is shown to be a very important energy transport mechanism in the atmosphere also. A physical explanation of why rising air cools and sinking air warms is given.

The nature of and rules which govern the emission of electromagnetic radiation are reviewed next. Students should find the discussion of sunburning and UVB radiation in this section interesting and relevant. The atmospheric greenhouse effect and the exchange of energy between the earth's surface, the atmosphere, and space are examined in detail. Students will see that, because the amounts of energy absorbed and emitted by the earth are in balance, the earth's average radiative equilibrium temperature varies little from year to year. Students should understand that the energy the earth absorbs from the sun consists primarily of short-wave radiation. Energy emitted by the earth is almost entirely in the form of infrared radiation. Selective absorbers in the atmosphere, such as water vapor and carbon dioxide, absorb some of the earth's infrared radiation and re-radiate a portion of it back to the surface. Because of this effect, the earth's average surface temperature is much higher than would otherwise be the case. A useful focus section describes this effect in the context of radiative equilibrium. Results from recent research relating to the effect of increasing concentrations of carbon dioxide and other greenhouse gases and the effects of clouds on the earth's energy balance are reviewed.

The final portion of the chapter describes the physical characteristics of the sun and the causes of the aurora.

Key Terms

energy
work
potential energy
kinetic energy
law of conservation
  of energy
first law of
  thermodynamics
temperature
heat
absolute zero
Kelvin scale
Fahrenheit scale
Celsius scale
internal energy
heat capacity
specific heat
calorie
latent heat
evaporation
condensation
sublimation
deposition
melting
freezing
sensible heat
conduction
heat (thermal
  conductivity)
convection

thermals
convective circulation
advection
parcel
radiant energy
  (radiation)
electromagnetic waves
wavelength
micrometer
photons
Stefan-Boltzmann Law
Wien's Law
longwave radiation
shortwave radiation
electromagnetic spectrum
visible region
visible light
ultraviolet (UV)
  radiation
infrared (IR)
  radiation
infrared sensors
UVA, UVB, UVC
  radiation
melanin
black body
radiative equilibrium
  temperature
selective absorbers
Kirchoff's Law

greenhouse effect
atmospheric greenhouse
  effect
atmospheric window
general circulation
  models (GCMs)
positive feedback
Earth Radiation Budget
  Experiment (ERBE)
negative feedback
free convection cells
solar constant
scattering
diffuse light
reflected (light)
albedo
photosphere
sunspots
faculae
chromosphere
corona
prominences
flares
magnetic storm
plasma
solar wind
magnetosphere
aurora borealis
aurora australis
airglow

Teaching Suggestions

1.   Heat a thin iron bar in a flame (from a Bunsen burner or a propane torch). Begin by holding the bar fairly close to the end of the bar. Students will see that heat is quickly conducted through the metal when the instructor is forced to move his or her grip down the bar. Repeat the demonstration with a piece of glass tubing or glass rod. Glass is a poor conductor and the instructor will be able to comfortably hold the glass just 2 or 3 inches from the tip. Ask the students if they believe energy is being transported away from the hot glass and if so, how? Without heat loss by conduction, the glass will get hotter than the iron bar and the tip should begin to glow red - a good demonstration of energy transport by radiation. Faint convection currents in the air can be made visible if the hot piece of glass is held between an overhead projector and the projection screen. Ask the students what they would do to quickly cool a hot object. Many will suggest blowing on it, an example of forced convection. Someone might suggest plunging the hot object into water. This makes for a satisfying end to the demonstration. Evaporating water can be seen and heard when the hot iron rod is put into the water (the glass will shatter if placed in the water).

The speed with which the rod is cooled is proof of the large amount of latent heat energy associated with changes of phase.

2.   Ask the students whether they believe water could be brought to a boil most rapidly in a covered or an uncovered pot. The question can be answered experimentally by filling two beakers with equal amounts of water and placing them on a single hot plate (to insure that energy is supplied to both at equal rates). It is a good idea to place boiling stones in the beakers to insure gentle boiling. Cover one of the beakers with a piece of foil. The covered pot will boil first. Explanation: a portion of the energy added to uncovered pot is used to evaporate water, not to increase the water's temperature.

3.   The concept of equilibria is sometimes difficult for students to grasp. Place a glass of water on a table top and ask the students whether they think the temperature of the water in the glass is warmer, cooler or the same as the surroundings. Many will say it is the same. Ask the students whether they think there is any energy flowing into or out of the glass. With some encouragement, they will recognize that the water is slowly evaporating and that this represents energy flow out of the glass. Energy flowing out of the glass will cause the water's temperature to decrease. Will the water just continue to get colder and colder until it freezes? No, as soon as the water's temperature drops below the temperature of the surroundings, heat will begin to flow into the water. The rate at which heat flows into the glass will depend on the temperature difference between the glass and the surroundings. The water temperature will decrease until energy flowing into the glass balances the loss due to evaporation.

4.   Use a lamp with a 150 Watt reflector bulb to help explain the concept of radiation intensity. Blind-fold a student and hold the lamp at various distances from the student's back. Ask the student to judge the distance of the bulb. Use the same lamp to illustrate the concepts of reflection, albedo, and absorption by measuring the amount of reflected light from various colored surfaces with a sensitive light meter. The reflectivity of natural surfaces outdoors could be measured or form the basis for a student or group project.

5   A 200 Watt clear light bulb connected to a dimmer switch can be used to illustrate how the temperature of an object affects the amount and type of radiation that the object will emit. Explain that passage of electricity through the resistive filament heats the filament. The filament's temperature will increase until it is able to emit energy at the same rate as it gains energy from the electric current. With the dimmer switch set low, the bulb can be made to glow a dull red. At low temperatures, the bulb emits low-intensity, longwave radiation. As the setting on the dimmer switch is increased, the color of the filament will turn orange, yellow and then white as increasing amounts of shortwave radiation are emitted. The intensity of the radiation will increase dramatically.

6.   Many students don't understand that a colored object appears that way because it reflects or scatters light of that color. The object isn't emitting visible light (ask the student whether they would see the object if all the lights in the room were turned off). Some students have the misconception that a green object reflects all colors but green. Similarly it is important that students understand that a red or green filter transmits red or green light. Put a red and a green (or blue) filter on an overhead projector and draw a hypothetical filter transmission curve. Put the two filters together and show that no light is transmitted. Ask the students what happens to the light that is not transmitted by the filter.

7.   Thought experiment to illustrate the magnitude of latent heat of evaporation/condensation: Ask students to think about taking a hot shower. Their body temperature is ~ 100°F; the water temperature is > 100°F; the air temperature in the room is ~75°F. Why, then, do you feel cold when you step, dripping

wet, out of the shower?

 8.   Use the Atmospheric Basics/Energy Balance section of the BlueSkies cdrom to step through the earth-atmosphere energy balance at various times of day, to demonstrate the magnitudes of the different energy budget components.

Student Projects

1.   Solar irradiance (energy per unit time per unit area) at the ground can be measured relatively easily. Begin with a rectangular piece of aluminum a few inches on a side and 3/8 or 1/2 inch thick. Drill a hole in one side so that a thermometer can be inserted into the middle of the block. Paint one of the two surfaces with flat black paint. Position the block in a piece of styrofoam insulation so that the painted surface faces outward and is flush with the styrofoam surface. Insert the thermometer into the side of the block. Orient the block so that the black surface is perpendicular to incident radiation from the sun. Note the time and measure the block temperature every 30 seconds for 10 to 15 minutes. When plotted on a graph, students should find that temperature, T, increases linearly with time, t. The slope of this portion of the graph can be used to infer the solar irradiance, S, using the following equation:

$$S = \frac{mass \times specific\,heat}{area} \times \frac{\Delta T}{\Delta t}$$

 2.   Use the Atmospheric Basics/Energy Balance section of the BlueSkies cdrom to compare the solar energy balance for Goodwin Creek, MS and Fort Peck, MT. What are the noontime albedos for each location? Why are they different? Which component of the albedo (earth's surface, clouds, or atmosphere) dominates in each case? Explain why.

 3.   Using the Atmospheric Basics/Energy Balance section of the BlueSkies cdrom, compare the values of the wintertime earth-atmosphere energy balance components for Penn State, PA and Desert Rock, NV. Explain any differences you find.

Answers to Questions for Thought

1.   The bridge will become icy first because it is able to lose heat energy over its entire surface; it cools on top, on the sides, and on the underside. The road, on the other hand, loses heat energy quickly, but only at its upper surface. Also, when the road begins to cool heat may flow up from warmer ground below.

2.   The branches cool rapidly by emitting infrared energy. The bare ground cools also, but it gains heat from the warmer soil below. Thus, the temperature of the bare ground may not drop below freezing and the freshly fallen snow will melt.

3.   These objects must be good emitters of radiation. Good emitters of radiation will cool to temperature less than that of the surrounding air. Energy lost by radiation is not quickly replaced by conduction. Air is a selective emitter of radiation and does not cool as rapidly as the ground.

4.   The ice can form when the air is <u>dry</u> and a <u>strong wind</u> blows over the water, causing rapid evaporation and cooling to the freezing point.

5.   Winter. Even though the oceans are cooler in winter than in summer, there is a greater temperature contrast between the oceans and the atmosphere in winter.

6.   In the form of electromagnetic radiation only.

7.   Ultraviolet radiation carries more energy per photon than does visible radiation.

8.   At a given distance from the large fire the energy received per unit area and per unit time is greater than the energy received at the same distance from the small fire.

9.   Without water vapor to absorb the earth's emitted infrared radiation, the earth will lose more heat.

10.   A plowed field. A plowed field is dark and has a low albedo - it is a poor reflector and a good absorber of sunlight. The snow surface has a high albedo and is a good reflector and poor absorber of sunlight.

11.   The low cloud absorbs energy emitted by the earth's surface and re-radiates infrared radiation back to the surface. A portion of the energy lost by the earth is returned.

12.   Removing the water vapor, because water vapor is a strong absorber of infrared radiation and atmospheric concentrations of H_2O are much higher than concentrations of CO_2.

13.   An increase in cloud cover would increase the earth-atmosphere albedo and, thus, less sunlight would reach the earth's surface. Depending on the height and thickness of the cloud cover, the clouds might absorb more infrared earth radiation and, thus, tend to strengthen the atmospheric greenhouse effect.

14.   This could happen in the upper atmosphere where the air is quite thin. Here the molecules move at average speeds proportional to a temperature of 1000 °C. However, few molecules would strike the thermometer and transfer heat to it. Consequently, the thermometer would lose energy much faster than it would gain energy. The thermometer would cool until it eventually registered a temperature near -273 °C.

15.   The energized particles from a large solar flare, that may produce auroral displays at lower latitudes, usually take a day or so to reach the earth's outer atmosphere.

16.   In Fig. 2.20 note that the aurora belt extends closer to Maine than to Washington state. The aurora belt circles the magnetic north pole, not the geographic north pole.

Answers to Problems and Exercises

1.   500 g x 600 cal/g = 300,000 calories
      300,000 cal/(100,000 g x 0.24 cal/gm °C) = 12.5 °C warmer

2.   Planet A, with the largest surface area, would be emitting the most radiation. The wavelength of maximum emission for both planets would be = 3000/1500 = 2 m.

3.   (a) the wavelength of maximum emission for Planet B would be 1 m.
(b) Near-infrared.
(c) Once its temperature is doubled, Planet B emits 8 times more energy per unit time than Planet A. Once its temperature doubles, Planet B would emit 16 times more energy per unit area of surface than Planet A (Stefan-Boltzmann law). Planet B has only half the total surface area that Planet A does however.

4.   Radiant energy $E = \sigma T^4$. Converting T from Fahrenheit to Kelvin gives T = [5/9 x (90-32) + 273] = 305.2 K. Using T = 305.2 K and σ=5.67 x 10^{-8} $W/m^2/K^4$, we find $E = 492$ W/m^2.

Multiple Choice Exam Questions

1.   Which of the following provides a measure of the average speed of air molecules:
a. pressure
b. temperature
c. density
d. heat

ANSWER: b

2.   A change of one degree on the Celsius scale is ____ a change of one degree on the Fahrenheit scale.
a. equal to
b. larger than
c. smaller than
d. is in the opposite direction of

ANSWER: b

3.   Which of the following is not considered a temperature scale?
a. Fahrenheit
b. Kelvin
c. Calorie
d. Celsius

ANSWER: c

4.   The temperature scale where 0° represents freezing and 100° boiling:
      a. Fahrenheit
      b. Celsius
      c. Kelvin
      d. absolute

ANSWER: b

5.   The temperature scale that sets freezing of pure water at 32° F:
      a. Kelvin
      b. Fahrenheit
      c. Celsius
      d. British

ANSWER: b

6.   If the temperature of the air is said to be at absolute zero, one might conclude that:
      a. the motion of the molecules is at a maximum
      b. the molecules are occupying a large volume
      c. the molecules contain a minimum amount of energy
      d. the temperature is 0° F
      e. the air temperature is 0° C

ANSWER: c

7.   In the Celsius temperature scale, what is the significance of the temperature increment of 1°C?
      a. it is the freezing point of water
      b. it is the boiling point of salt water
      c. it is one-tenth of the interval between the freezing point and the boiling point of water
      d. it is one-tenth of the interval between the freezing point and the boiling point of salt water
      e. it is 1/100 of the interval between the freezing point and the boiling point of water

ANSWER: e

8.   Energy of motion is also known as:
      a. dynamic energy
      b. kinetic energy
      c. sensible heat energy
      d. static energy
      e. latent heat energy

ANSWER: b

9. Heat is energy in the process of being transferred from:
    a. hot objects to cold objects
    b. low pressure to high pressure
    c. cold objects to hot objects
    d. high pressure to low pressure
    e. regions of low density toward regions of high density

ANSWER: a

10. The heat energy released when water vapor changes to a liquid is called:
    a. latent heat of evaporation
    b. latent heat of fusion
    c. latent heat of fission
    d. latent heat of condensation

ANSWER: d

11. The change of state of ice into water vapor is known as:
    a. deposition
    b. sublimation
    c. melting
    d. condensation
    e. crystallization

ANSWER: b

12. When water changes from a liquid to a vapor, we call this process:
    a. freezing
    b. condensation
    c. sublimation
    d. deposition
    e. evaporation

ANSWER: e

13. This is released as sensible heat during the formation of clouds:
    a. potential energy
    b. longwave radiation
    c. latent heat
    d. shortwave radiation
    e. kinetic energy

ANSWER: c

14. The cold feeling that you experience after leaving a swimming pool on a hot, dry, summer day represents heat transport by:
    a. conduction
    b. convection
    c. radiation
    d. latent heat

ANSWER: d

15. The term "latent" means:
    a. late
    b. hot
    c. light
    d. hidden
    e. dense

ANSWER: d

16. The processes of condensation and freezing:
    a. both release sensible heat into the environment
    b. both absorb sensible heat from the environment
    c. do not affect the temperature of their surroundings
    d. do not involve energy transport

ANSWER: a

17. The transfer of heat by molecule-to-molecule contact:
    a. conduction
    b. convection
    c. radiation
    d. ultrasonic

ANSWER: a

18. Which of the following is the poorest conductor of heat?
    a. still air
    b. water
    c. ice
    d. snow
    e. soil

ANSWER: a

19.   The horizontal transport of any atmospheric property by the wind is called:
       a. advection
       b. radiation
       c. conduction
       d. latent heat
       e. reflection

ANSWER: a

20.   A heat transfer process in the atmosphere that depends upon the movement of air is:
       a. conduction
       b. absorption
       c. reflection
       d. convection
       e. radiation

ANSWER: d

21.   The amount of heat energy required to bring about a small change in temperature is called the
       a. radiative equilibrium
       b. dead heat
       c. specific heat
       d. latent heat

ANSWER: c

22.   Snow will usually <u>melt</u> on the roof of a home that is a:
       a. good radiator of heat
       b. good conductor of heat
       c. poor radiator of heat
       d. poor conductor of heat

ANSWER: b

23.   Rising air cools by this process:
       a. expansion
       b. evaporation
       c. compression
       d. condensation

ANSWER: a

24.   The temperature of a rising air parcel:
        a. always cools due to expansion
        b. always warms due to expansion
        c. always cools due to compression
        d. always warms due to compression
        e. remains constant

ANSWER: a

25.   The proper order from shortest to longest wavelength is:
        a. visible, infrared, ultraviolet
        b. infrared, visible, ultraviolet
        c. ultraviolet, visible, infrared
        d. visible, ultraviolet, infrared
        e. ultraviolet, infrared, visible

ANSWER: c

26.   Sinking air warms by this process:
        a. compression
        b. expansion
        c. condensation
        d. friction

ANSWER: a

27.   Heat transferred outward from the surface of the moon can take place by:
        a. convection
        b. conduction
        c. latent heat
        d. radiation

ANSWER: d

28.   How do red and blue light differ?
        a. blue light has a higher speed of propagation
        b. the wavelength of red light is longer
        c. red light has a higher intensity
        d. red and blue light have different directions of polarization

ANSWER: b

29.   If the average temperature of the sun increased, the wavelength of peak solar emission would:
        a. shift to a shorter wavelength
        b. shift to a longer wavelength
        c. remain the same
        d. impossible to tell from given information

ANSWER: a

30.   Solar radiation reaches the earth's surface as:
        a. visible radiation only
        b. ultraviolet radiation only
        c. infrared radiation only
        d. visible and infrared radiation only
        e. ultraviolet, visible, and infrared radiation

ANSWER: e

31.   Electromagnetic radiation with wavelengths between 0.4 and 0.7 micrometers is called:
        a. ultraviolet light
        b. visible light
        c. infrared light
        d. microwaves

ANSWER: b

32.   The sun emits a maximum amount of radiation at wavelengths near ___, while the earth emits maximum radiation near wavelengths of ___.
        a. 0.5 micrometers, 30 micrometers
        b. 0.5 micrometers, 10 micrometers
        c. 10 micrometers, 30 micrometers
        d. 1 micrometer, 10 micrometers

ANSWER: b

33.   The blueness of the sky is mainly due to:
        a. the scattering of sunlight by air molecules
        b. the presence of water vapor
        c. absorption of blue light by the air
        d. emission of blue light by the atmosphere

ANSWER: a

34. Which of the following determine the kind (wavelength) and amount of radiation that an object emits?
    a. temperature
    b. thermal conductivity
    c. density
    d. latent heat

ANSWER: a

35. Often before sunrise on a clear, calm, cold morning, ice (frost) can be seen on the tops of parked cars, even when the air temperature is above freezing. This condition happens because the tops of the cars are cooling by:
    a. conduction
    b. convection
    c. latent heat
    d. radiation

ANSWER: d

36. One micrometer is a unit of length equal to:
    a. one million meters
    b. one millionth of a meter
    c. one tenth of a millimeter
    d. one thousandth of a meter

ANSWER: b

37. Evaporation is a _____ process.
    a. cooling
    b. heating
    c. can't tell - it depends on the temperature
    d. both a and c

ANSWER: a

38. Which of the following represents the smallest unit of length?
    a. mile
    b. centimeter
    c. meter
    d. micrometer
    e. inch

ANSWER: d

39. Which of the following has a wavelength shorter than that of violet light?
    a. green light
    b. blue light
    c. infrared radiation
    d. red light
    e. ultraviolet radiation

ANSWER: e

40. If the absolute temperature of an object doubles, the maximum energy emitted goes up by a factor of:
    a. 2
    b. 4
    c. 8
    d. 16
    e. 32

ANSWER: d

41. At which temperature would the earth be radiating energy at the greatest rate or intensity?
    a. -5° F
    b. -40° F
    c. 60° F
    d. 32° F
    e. 105° F

ANSWER: e

42. How much radiant energy will an object emit if its temperature is at absolute zero?
    a. the maximum theoretical amount
    b. none
    c. the same as it would at any other temperature
    d. depends on the chemical composition of the object

ANSWER: b

43. Most of the radiation emitted by a human body is in the form of:
    a. ultraviolet radiation and is invisible
    b. visible radiation but is too weak to be visible
    c. infrared radiation and is invisible
    d. humans do not emit electromagnetic radiation

ANSWER: c

44. Clouds *never* form by
    a. sublimation
    b. condensation
    c. evaporation
    d. deposition
    e. both a and c

ANSWER: e

45. The sun emits its greatest intensity of radiation in:
    a. the visible portion of the spectrum
    b. the infrared portion of the spectrum
    c. the ultraviolet portion of the spectrum
    d. the x-ray portion of the spectrum

ANSWER: a

46. Air that rises always
    a. contracts and warms
    b. contracts and cools
    c. expands and cools
    d. expands and warms

ANSWER: c

47. The earth's radiation is often referred to as __ radiation, while the sun's radiation is often referred to as __ radiation.
    a. shortwave, longwave
    b. shortwave, shortwave
    c. longwave, shortwave
    d. longwave, longwave

ANSWER: c

48. If the earth's average surface temperature were to increase, the amount of radiation emitted from the earth's surface would _____ and the wavelength of peak emission would shift toward _____ wavelengths.
    a. increase, shorter
    b. increase, longer
    c. decrease, shorter
    d. decrease, longer

ANSWER: a

49. A football field is about _____ micrometers long.
    a. 10^{-8}
    b. 10^{-6}
    c. 10^{6}
    d. 10^{8}

ANSWER: d

50. The earth emits radiation with greatest intensity at:
    a. infrared wavelengths
    b. radio wavelengths
    c. visible wavelengths
    d. ultraviolet wavelengths

ANSWER: a

51. "A good absorber of a given wavelength of radiation is also a good emitter of that wavelength."
    This is a statement of:
    a. Stefan-Boltzmann's law
    b. Wien's Law
    c. Kirchoff's Law
    d. the First Law of Thermodynamics
    e. the Law of Relativity

ANSWER: c

52. Which principle best describes why holes develop in snow around tree trunks?
    a. snow is a good absorber of infrared energy
    b. snow is a good emitter of infrared energy
    c. snow is a poor reflector of visible light
    d. snow is a poor absorber of visible light
    e. snow is a poor absorber of ultraviolet light

ANSWER: a

53. Which of the following statements is not correct?
    a. calm, cloudy nights are usually warmer than calm, clear nights
    b. each year the earth's surface radiates away more energy than it receives from the sun
    c. the horizontal transport of heat by the wind is called advection
    d. good absorbers of radiation are usually poor emitters of radiation

ANSWER: d

54.   Without the atmospheric greenhouse effect, the average surface temperature would be:
        a. higher than at present
        b. lower than at present
        c. the same as it is now
        d. much more variable than it is now

ANSWER: b

55.   The earth's atmospheric window is in the:
        a. ultraviolet region
        b. visible region
        c. infrared region
        d. polar regions

ANSWER: c

56.   The atmospheric greenhouse effect is produced mainly by the:
        a. absorption and re-emission of visible light by the atmosphere
        b. absorption and re-emission of ultraviolet radiation by the atmosphere
        c. absorption and re-emission of infrared radiation by the atmosphere
        d. absorption and re-emission of visible light by clouds
        e. absorption and re-emission of visible light by the ground

ANSWER: c

57.   Suppose last night was clear and calm. Tonight low clouds will be present. From this you would
        conclude that tonight's minimum temperature will be:
        a. higher than last night's minimum temperature
        b. lower than last night's minimum temperature
        c. the same as last night's minimum temperature
        d. above freezing

ANSWER: a

58.   Which of the following is known primarily as a selective absorber of ultraviolet radiation?
        a. carbon dioxide
        b. ozone
        c. water vapor
        d. clouds

ANSWER: b

59. Low clouds retard surface cooling at night better than clear skies because:
   a. the clouds absorb and radiate infrared energy back to earth
   b. the water droplets in the clouds reflect infrared energy back to earth
   c. the clouds start convection currents between them
   d. the clouds are better conductors of heat than is the clear night air
   e. the formation of the clouds releases latent heat energy

ANSWER: a

60. At night, low clouds:
   a. enhance the atmospheric greenhouse effect
   b. weaken the atmospheric greenhouse effect
   c. are often caused by the atmospheric greenhouse effect
   d. have no effect on the atmospheric greenhouse effect

ANSWER: a

61. Which of the following gases are mainly responsible for the atmospheric greenhouse effect in the earth's atmosphere?
   a. oxygen and nitrogen
   b. nitrogen and carbon dioxide
   c. ozone and oxygen
   d. water vapor and carbon dioxide

ANSWER: d

62. Of the gases listed below, which is <u>not</u> believed to be responsible for enhancing the earth's greenhouse effect?
   a. chlorofluorocarbons (CFCs)
   b. molecular oxygen (O_2)
   c. nitrous oxide (N_2O)
   d. carbon dioxide (CO_2)
   e. methane (CH_4)

ANSWER: b

63. The combined albedo of the earth and the atmosphere is approximately:
   a. 4%
   b. 10%
   c. 30%
   d. 50%
   e. 90%

ANSWER: c

64. According to the Stefan-Boltzmann law, the radiative energy emitted by one square meter of an object is equal to a constant multiplied by its temperature raised to the _____ power.
    a. negative third
    b. zeroeth
    c. fourth
    d. tenth

ANSWER: c

65. The albedo of the moon is 7%. This means that:
    a. 7% of the sunlight striking the moon is reflected
    b. 7% of the sunlight striking the moon is absorbed
    c. the moon emits only 7% as much energy as it absorbs from the sun
    d. 93% of the sunlight striking the moon is reflected

ANSWER: a

66. If the present concentration of CO_2 doubles in 100 years, and climate models predict that for the earth's average temperature to rise 5° C, what gas must also increase in concentration?
    a. nitrogen
    b. oxygen
    c. methane
    d. water vapor

ANSWER: d

67. The albedo of the earth's surface is only about 4%, yet the combined albedo of the earth and the atmosphere is about 30%. Which set of conditions below best explains why this is so?
    a. high albedo of clouds, low albedo of water
    b. high albedo of clouds, high albedo of water
    c. low albedo of clouds, low albedo of water
    d. low albedo of clouds, high albedo of water

ANSWER: a

68. According to Wein's displacement law, the wavelength at which maximum radiation occurs
    a. is inversely proportional to the temperature
    b. is proportional to the temperature
    c. is inversely proportional to the pressure
    d. is proportional to the pressure

ANSWER: a

69.   Clouds ___ infrared radiation and ___ visible radiation.
        a. absorb, absorb
        b. absorb, reflect
        c. reflect, reflect
        d. reflect, absorb

ANSWER: b

70.   An increase in albedo would be accompanied by ___ in radiative equilibrium temperature.
        a. an increase
        b. a decrease
        c. no change
        d. unstable oscillations

ANSWER: b

71.   On the average, about what percentage of the solar energy that strikes the outer atmosphere eventually reaches the earth's surface?
        a. 5%
        b. 15%
        c. 30%
        d. 50%
        e. 70%

ANSWER: d

72.   If the amount of energy lost by the earth to space each year were not approximately equal to that received:
        a. the atmosphere's average temperature would change
        b. the length of the year would change
        c. the sun's output would change
        d. the mass of the atmosphere would change

ANSWER: a

73.   If the sun suddenly began emitting more energy, the earth's radiative equilibrium temperature would:
        a. increase
        b. decrease
        c. remain the same
        d. begin to oscillate

ANSWER: a

74. Sunlight that bounces off a surface is said to be ___ from the surface.
    a. radiated
    b. absorbed
    c. emitted
    d. reflected

ANSWER: d

75. The major process that warms the lower <u>atmosphere</u> is:
    a. the release of latent heat during condensation
    b. conduction of heat upward from the surface
    c. absorption of infrared radiation
    d. direct absorption of sunlight by the atmosphere

ANSWER: c

76. Atmospheric concentrations of N_2O and CH_4 contribute _____ to the earth-atmosphere albedo.
    a. significantly
    b. little

ANSWER: b

77. The atmosphere near the earth's surface is "heated from below." Which of the following in <u>not</u> responsible for the heating?
    a. conduction of heat upward from a hot surface
    b. convection from a hot surface
    c. absorption of infrared energy that has been radiated from the surface
    d. heat energy from the earth's interior

ANSWER: d

78. The earth's radiative equilibrium temperature is:
    a. the temperature at which the earth is absorbing solar radiation and emitting infrared radiation at equal rates
    b. the temperature at which the earth is radiating energy at maximum intensity
    c. the average temperature the earth must maintain to prevent the oceans from freezing solid
    d. the temperature at which rates of evaporation and condensation on the earth are in balance

ANSWER: a

79.	In 2002, about _____ of the air near the surface is CO_2.
	a. 40%
	b. 4%
	c. 0.4%
	d. 0.04%

ANSWER: d

80.	Charged particles from the sun that travel through space at high speeds are called:
	a. radiation
	b. the aurora
	c. solar wind
	d. solar flares

ANSWER: c

81.	In the earth's upper atmosphere, visible light given off by excited atoms and molecules produces:
	a. flares
	b. the solar wind
	c. the aurora
	d. prominences

ANSWER: c

82.	The aurora is produced by:
	a. reflections of sunlight by polar ice fields
	b. fast-moving charged particles colliding with air molecules
	c. burning oxygen caused by the intense sunlight at high altitude
	d. the combination of molecular and atomic oxygen to form ozone
	e. scattering of sunlight in the upper atmosphere

ANSWER: b

83.	On a clear night, the best place to see the aurora would be:
	a. at the magnetic north pole
	b. northern Maine
	c. northern Washington
	d. Colorado
	e. Alaska

ANSWER: e

84. The luminous surface of the sun is known as the:
   a. chromosphere
   b. thermosphere
   c. corona
   d. photosphere
   e. exosphere

ANSWER: d

85. Sunspots:
   a. appear darker than the rest of the sun's surface
   b. are cooler regions on the sun's surface
   c. are located in regions of strong magnetic fields
   d. reach a maximum on the sun approximately every 11 years
   e. all of the above

ANSWER: e

86. The aurora are seen:
   a. in the Northern Hemisphere only
   b. in the Southern Hemisphere only
   c. in both the Northern and Southern Hemispheres at high latitudes
   d. in both the Northern and Southern Hemispheres near the equator

ANSWER: c

87. Suppose you are outside in very cold temperatures, wearing a winter coat that is quite effective at keeping you warm. Which of the following is true?
   a. The coat is the source of the heat that keeps you warm.
   b. Your body generates the heat that keeps you warm.
   c. The coat prevents your body's heat from escaping to the surrounding air.
   d. both (a) and (c) are true.
   e. both (b) and (c) are true.

ANSWER: e

Essay Exam Questions

1. What is meant by the term "positive feedback?" What role could positive feedback play in the atmospheric greenhouse effect? Would this enhance or reduce global warming? Can you think of any "negative feedback" mechanisms?

2. In the discussion of the earth's annual energy balance we saw that the earth absorbed approximately 51 units of solar energy but emitted 117 units of infrared energy. What prevents the earth from getting colder and colder?

3. Will a rising parcel of air always expand? Why? Does this expansion cause the air temperature to increase or decrease? Why?

4. Explain how energy in the form of sunlight absorbed at the ground could be transferred upward in the atmosphere in the form of latent heat. How or when is the latent heat energy released in the air above the ground?

5. Describe and give examples of the various ways that heat can be transported in the atmosphere.

6. Describe the atmospheric greenhouse effect. Is there any difference between the way the atmospheric greenhouse effect works on a clear night and on a cloudy night?

7. Several of the planets in our solar system are further from the sun and cooler than the earth. Do they emit electromagnetic radiation? Why are we able to see the planets in the sky at night?

8. How could increased cloud cover cause an increase in the average surface temperature? How could increased cloudiness cause a decrease in average surface temperatures?

9. When you remove a cold beverage from a refrigerator in a humid room, water vapor will condense on the sides of the container. Would this act to warm or cool the beverage, or would the condensation have no effect on the beverage's temperature?

10. Imagine that the temperature of the sun were to change. Describe or discuss some of the effects that this might have on the earth's energy budget and the earth's climate.

11. Many automobile engines are cooled by water which flows in a closed circuit through the engine block and the car's radiator. How many different heat transport processes do you find in operation here?

12. Many people will blow on a bowl of hot soup to try to cool it. In your view, what are the two most important heat transport processes being used to cool the soup?

13. In what ways is the atmospheric greenhouse different from an agricultural greenhouse?

14. What are the other factors, besides increasing CO_2 concentrations, that affect global warming?

Chapter 3
Seasonal and Daily Temperatures

Summary

This chapter begins with an explanation of the causes of the earth's changing seasons. A full year's cycle of the seasons is described for the Northern Hemisphere. Daily, seasonal, and geographic variations in temperature have important practical and economic implications and are examined next. Daily temperature is controlled by incoming energy, primarily from the sun, and outgoing energy from the earth's surface. While energy from the sun is generally most intense at noon, daytime temperatures continue to rise into the afternoon as long as energy input exceeds output. Because most of the incident sunlight is first absorbed at the ground and then transported up into the atmosphere, large temperature gradients can develop between the ground and the air just above, especially under calm wind conditions. At night, the ground cools more rapidly than the air above and a radiation inversion will often form. Students will see that with an understanding of the factors which promote the formation of an inversion layer, it is often possible for farmers and growers to reduce the severity of a nighttime inversion and to protect cold-sensitive plants and trees. The psychological effects of seasonal change are also briefly mentioned.

Temperature varies considerably on a geographical scale and mean and record temperatures observed throughout the world are summarized. The main factors that affect the range of temperatures at different locations around the world are latitude, elevation and proximity to land, water or ocean currents. Variables that can be used to characterize the climate of different regions, such as mean daily temperature, mean annual temperature and annual range of temperature, are presented. Additional parameters are described, such as the number of heating, cooling and growing degree-days, which can be used to estimate a region's heating or cooling needs. The updated wind-chill index is also presented. A focus section is included that describes the concept of 'normal' temperatures.

The chapter concludes with an examination of how the human body's perception of temperature is influenced by atmospheric conditions and discusses the different types of thermometers and instruments that are used to measure temperature.

Key Terms

perihelion
aphelion
summer solstice
Arctic Circle
insolation
Tropic of Cancer
autumnal equinox
Indian summer
chlorophylls
winter solstice
Tropic of Capricorn
cold wave
vernal equinox
lag in seasonal
  temperature
forced convection
radiational cooling
radiation inversion
nocturnal inversion

thermal belts
orchard heaters
smudge pots
wind machines
freeze
advection frost
frost ban
controls of temperature
isotherms
specific heat
daily (diurnal) range
  of temperature
mean daily temperature
annual range
  of temperature
mean (average) annual
temperature
heating degree-day
cooling degree-day

growing degree-days
sensible temperature
wind-chill factor
frostbite
hypothermia
liquid-in-glass
  thermometers
maximum thermometer
minimum thermometer
electrical thermometer
electrical resistance
  thermometer
thermistors
thermocouple
infrared sensors
radiometers
bimetallic thermometer
thermograph
instrument shelter

Teaching Suggestions

1.   Begin the lecture by drawing an ellipse on the blackboard with the sun positioned much closer to one end of the ellipse. On the other end of the ellipse, closest to the sun, make a dot for the earth and label it, "January and winter." Then label the other end "July and summer." Act confused and ask, "Wait a minute, is that correct?" Usually this is enough to start an interesting discussion on what causes the seasons.

2.   Explain the seasons by shining a fairly broad, collimated beam of light onto a globe in a darkened room. Begin by showing the earth with no tilt, then increase the tilt to 23.5°. Finally increase the tilt to 45°. Explain how the change in tilt would influence the average temperature measured in July and January in the Northern Hemisphere. Discuss the design of houses to take advantages of the earth-sun geometry: Where would you plant shade trees? Would you want them to be deciduous or evergreen?

        Using the globe or drawings, the students should understand whether they would need to look to the south or north of overhead to see the sun a noon from different locations on the earth at different times of year. They should also understand whether it is necessary to look east, northeast, or southeast to see the sun rise.

3.   The attenuation of light as is passes through a scattering medium can be demonstrated by placing a photodetector at one end of an aquarium full of water and a light source at the other end. Then begin to add milk in small, but reproducible amounts. The signal at the detector will decay exponentially with the amount of milk added. The decay will depart from the exponential law when enough milk is added that appreciable multiple scattering begins to occur. The effect of an absorbing medium can be demonstrated if India ink is used in place of milk.

4.   A photodetector can be placed on a flat board that is then oriented perpendicularly to light rays coming from a source placed further away. As the detector is tilted with respect to the light source, the signal will decrease. If the tilt angle is measured, the photodetector signal will be seen to obey the cosine law.

5.   Make an inversion. Fill a plastic tub with crushed ice. Attach 3 or 4 thermometers to a vertical ringstand such that the thermometers are several inches apart and the lowest thermometer is about 1 inch above (but not touching) the ice. Cover the thermometers with a large glass bell jar.

Make a graph of the change in temperature with time for each thermometer. At first, the readings will be isothermal, but eventually the coldest air will be observed just above the ice and a strong inversion will have formed.

This demonstration can be applied to important concepts presented in this and the previous chapter. Explain why the air cools faster near the ice and relate this to the formation of a radiation inversion. Ask what ingredients are necessary to maintain the inversion. Explain how the inversion would be destroyed. Ask what kind of weather conditions would be necessary to get an inversion.

6.   Pass around a piece of metal, a piece of wood, and a piece of styrofoam (all three could be attached to a larger piece of plywood, perhaps). Explain that all three objects have been sitting in the classroom and have the same temperature. Ask the students whether all three objects <u>feel</u> like they have the same temperature. This is a good demonstration that our perception of temperature is often a better indication of how quickly our body is losing heat rather than absolute temperature and leads into a discussion of the wind-chill effect.

Student Projects

1.   Have students measure the temperature of a variety of different surfaces on a sunny afternoon. Are there differences in the temperature values? Ask the students to explain what causes the differences.

2.   Have several students, who live in different parts of a city or town, make simultaneous early morning temperature measurements.   Are there appreciable differences?   Can these differences be attributed to topography? Is there evidence of a "thermal belt" in their city?

3.   Have the students plot an early morning and late afternoon sounding. Is a radiation inversion visible on the morning sounding? If so, how deep is the inversion layer? Did the students observe any visual evidence of a strong radiation inversion that morning? How had the sounding changed by that afternoon? How much of a change was observed at the ground and at different levels above the ground? (Tabulated data may use pressure as the vertical coordinate; in this case it is probably sufficient to assume a 1 mb decrease per 10 meters)

 Blue Skies   4.   Using the Atmospheric Basics/Layers of the Atmosphere section of the BlueSkies cdrom, find an upper-air site in the continental U.S. that has an inversion. What weather conditions do you think are causing this inversion?

Blue Skies 5.   Use the Atmospheric Basics/Energy Balance section of the BlueSkies cdrom to explore the model of energy exchange between the surface and the atmosphere. Use the time slider to step the model through three-hour increments. For each time step, note the values of energy flow to and from the earth's surface, and plot the total gains and losses versus time. Based on this information, when would you expect the temperature to reach its peak for the day?

Blue Skies 6.   Using the Weather Forecasting/Forecasting section of the BlueSkies cdrom, click on World Weather to show current temperatures in different hemispheres (that are experiencing different seasons). Do the temperatures you find match your expectations? Why or why not?

Answers to Questions for Thought

1.   The diagram should look similar to Fig. 3.10 (the sun rises in the SE in winter and in the NE in summer) except that a house with a bedroom window facing south should be at the center of the illustration.

2.   (a) In middle and high latitudes the summer sun would be lower and the winter sun higher. This should produce cooler summers and milder winters than now. (b) In middle latitudes the summer sun would be higher and the winter sun lower. This should produce warmer summers and colder winters than now.

3.   The most solar radiation per day would be received at the top of the earth's atmosphere near latitude 80° N or 85° N (at the North Pole on June 21st). Maximum radiation in one day would be found at the earth's surface near latitude 30° N. At 30° N, the days are shorter but there are fewer clouds so less sunlight is scattered and reflected, and more sunlight reaches the surface.

4.   North.

5.   Painting the north-facing wall a light color will reflect summer sunlight, while painting the south-facing wall a dark color will absorb winter sunlight.

6.   Because water heats and cools more slowly than land, the lag in daily temperature over water is usually much larger than that experienced over land.

7.   On a small island near the equator. Here, there would be an extremely small annual temperature range.

8.   San Francisco's average summer temperature is much lower than Richmond's average summer temperature. Consequently, more heating degree-days are computed at San Francisco during the summer than are computed at Richmond for the same time period.

9.   The highest daily range probably occurred in New Mexico where relatively dry summer air allows for large daily temperature ranges. The lowest daily range more than likely occurred in New Jersey where the humid summer air keeps the daily temperature range small.

10.   Because the maximum daily temperature usually occurs in the afternoon. The minimum daily temperature is unlikely to exceed this value.

11.   The clouds absorb some of the infrared radiation given off by the earth, thereby increasing their temperature. The warming clouds radiate energy downward, transferring the heat to the earth's surface.

12.   The latitude at which the noontime sun is directly overhead always lies between 23.5°S and 23.5°N. Most northern hemisphere locations are north of this range. This means that a south-facing slope will be facing the sun for much of the day, thus warming up more than a north-facing slope which is more often in shadow. The increased daytime temperatures of the south-facing slope (relative to the north-facing slope) increase the diurnal temperature range, since nighttime temperatures of the two slopes are likely to be similar.

13.   Two reasons: (1) the sun is quite low on the horizon, thereby providing a small insolation, and (2) much of the Earth's surface in the polar regions is white, resulting in a very high albedo.

Answers to Problems and Exercises

2.   (a) 47° change in elevation between winter and summer solstice.
        47° change in about 182 days or about 0.26° per day.
        (b) The same amount of change takes place (0.26°/day). However, above
66.5 °N the angle between winter and summer solstice ranges from 47° at 66.5° N to 23.5° at the North Pole.

3.   (a) The sun moves (as we observe it) about 0.26°/day. Therefore to move to a position of 10° N it takes $10°/(0.26°$ day^{-1}) or about 38 days before the autumnal equinox and 38 days after the vernal equinox, or approximately August 16 and April 30.
        (b) $15°/(0.26°$ day^{-1}) equals about 58 days. The sun will be overhead at noon at 15° S on approximately November 20 and January 23.

5.   In 40 days, or about June 9.

6.   The wind-chill equivalent temperature is -21 °F.

Multiple Choice Exam Questions

1.   During the winter in the Northern Hemisphere, the "land of the midnight sun" would be found:
      a. at high latitudes
      b. at middle latitudes
      c. near the equator
      d. in the desert southwest
      e. on the West Coast

ANSWER: a

2.   During the course of a year the sun will disappear from view near the <u>North Pole</u> on what date?
      a. June 21
      b. September 23
      c. December 23
      d. January 1
      e. March 21

ANSWER: b

3.   In the Northern Hemisphere, this day has the fewest hours of daylight:
      a. summer solstice
      b. winter solstice
      c. vernal equinox
      d. autumnal equinox

ANSWER: b

4.   During an equinox:
      a. the days and nights are of equal length except at the poles
      b. at noon the sun is overhead at the equator
      c. the earth is not tilted toward nor away from the sun
      d. all of the above

ANSWER: d

5.   Indian summer would most likely occur during the month of:
      a. October
      b. December
      c. June
      d. August

ANSWER: a

6.   Which of the following <u>best</u> describes the weather conditions necessary to bring Indian summer weather to the eastern half of the United States?
      a. a cold front moving off the New Jersey coast
      b. a strong slow-moving low pressure area just east of Virginia
      c. a strong slow-moving high pressure area off the southeast coast
      d. a strong fast-moving low pressure area over Georgia
      e. a cold front that stretches from South Carolina to Texas

ANSWER: c

7.   During the winter solstice in the Northern Hemisphere:
        a. astronomical winter begins in the Northern Hemisphere
        b. the noon sun is overhead at 23.5° S latitude
        c. at middle latitudes in the Northern Hemisphere, this marks the longest night of
           the year
        d. all of the above

ANSWER: d

8.   Which latitude below would experience the fewest hours of daylight on Dec. 22?
        a. 60° S
        b. 20° S
        c. 0° (Equator)
        d. 20° N
        e. 60° N

ANSWER: e

9.   Considering each hemisphere as a whole, seasonal temperature variations in the Southern
        Hemisphere are _____ those in the Northern Hemisphere.
        a. greater than
        b. about the same as
        c. less than

ANSWER: c

10.   Where are the days and nights of equal length all year long?
        a. at 66.5°
        b. nowhere
        c. at 23.5°
        d. at the Equator

ANSWER: d

11.   In the middle latitudes of the Northern Hemisphere on June 22, the sun:
        a. rises in the east and sets in the west
        b. rises in the southeast and sets in the southwest
        c. rises in the northeast and sets in the northwest
        d. rises in the northeast and sets in the southwest
        e. rises in the southeast and sets in the northwest

ANSWER: c

12. Which of the following helps to explain why even though northern latitudes experience 24 hours of sunlight on June 22, they are not warmer than latitudes further south?
    a. solar energy is spread over a larger area in northern latitudes
    b. some of the sun's energy is reflected by snow and ice in the northern latitudes
    c. increased cloud cover reflects solar energy in the northern latitudes
    d. solar energy is used to melt frozen soil in the northern latitudes
    e. all of the above

ANSWER: e

13. Treatment for Seasonal Affective Disorder generally involves increased exposure to
    a. sunlight
    b. carbon dioxide
    c. water vapor

ANSWER: a

14. The sun is directly overhead at Mexico City (latitude 19°N):
    a. once a year
    b. twice a year
    c. four times a year
    d. never

ANSWER: b

15. The north-facing side of a hill in a mountainous region in the US tends to:
    a. receive less sunlight during a year than the south-facing side
    b. grow a variety of trees that are typically observed at higher elevation
    c. be a better location for a ski run than the south-facing side
    d. have snow on the ground for a longer period of time in winter compared to the south-facing side
    e. all of the above

ANSWER: e

16. On what day would you expect the sun to be overhead at Lima, Peru (latitude 12° S)?
    a. August 15
    b. December 22
    c. February 4
    d. March 10
    e. April 21

ANSWER: c

17.   The maximum in daytime surface temperature typically occurs _____ the earth receives its most intense solar radiation.
      a. before
      b. after
      c. exactly when

ANSWER: b

18.   Radiational cooling typically occurs
      a. during the afternoon
      b. at night
      c. during the late morning

ANSWER: b

19.   The strongest radiation inversions occur when
      a. skies are overcast
      b. skies are partly cloudy
      c. skies are clear
      d. precipitation is falling

ANSWER: c

20.   When it is January and winter in the Northern Hemisphere, it is ___ and ___ in the Southern Hemisphere.
      a. January and summer
      b. January and winter
      c. July and winter
      d. July and summer

ANSWER: a

21.   The most important reason why summers in the Southern Hemisphere are not warmer than summers in the Northern Hemisphere is that:
      a. the earth is closer to the sun in January
      b. the earth is farther from the sun in July
      c. over 80% of the Southern Hemisphere is covered with water
      d. the sun's energy is less intense in the Southern Hemisphere

ANSWER: c

22.   For maximum winter warmth, in the Northern Hemisphere, large windows in a house should face:
      a. north
      b. south
      c. east
      d. west

ANSWER: b

23. Thermal belts are usually found
    a. on valley floors
    b. on hillsides
    c. on mountain tops

ANSWER: b

24. To protect fruit trees from frost, it is important to keep the air as still as possible.
    a. true
    b. false

ANSWER: b

25. During a radiation inversion, wind machines
    a. bring warm air down toward the surface
    b. lift cool, surface air to higher altitudes
    c. mix the air near the ground
    d. all of the above

ANSWER: d

26. The main reason(s) for warm summers in middle latitudes is that:
    a. the earth is closer to the sun in summer
    b. the sun is higher in the sky and we receive more direct solar radiation
    c. the days are longer
    d. all of the above
    e. only (b) and (c) are correct

ANSWER: e

27. Our seasons are caused by:
    a. the changing distance between the earth and the sun
    b. the angle at which sunlight reaches the earth
    c. the length of the daylight hours
    d. all of the above
    e. only (b) and (c) are correct

ANSWER: e

28. Incoming solar radiation in middle latitudes is less in winter than in summer because:
    a. the sun's rays slant more and spread their energy over a larger area.
    b. the sun's rays are weakened by passing through a greater thickness of
       atmosphere
    c. the cold dense air lowers the intensity of the sun's rays
    d. all of the above
    e. only (a) and (b) are correct

ANSWER: e

29. Using sprinklers to prevent crap damage from cold air works best when
   a. the air is fairly humid
   b. the air is fairly dry

ANSWER: a

30. At the North Pole the sun will rise above the horizon on __ and set below the horizon on __.
   a. June 22, September 23
   b. September 23, December 22
   c. March 21, September 23
   d. June 22, December 22
   e. March 21, December 22

ANSWER: c

31. The earth is tilted at an angle of $23.5°$ with respect to the plane of its orbit around the sun. If the amount of tilt were <u>increased</u> to $40°$, we would expect in middle latitudes:
   a. hotter summers and colder winters than at present
   b. cooler summers and milder winters than at present
   c. hotter summers and milder winters than at present
   d. cooler summers and colder winters than at present
   e. no appreciable change from present conditions

ANSWER: a

32. The latitude at which there is a yearly balance between incoming and outgoing radiation is about:
   a. $0°$
   b. $23.5°$
   c. $37°$
   d. $66.5°$
   e. $90°$

ANSWER: c

33. Although the polar regions radiate away more heat energy than they receive by insolation in the course of a year, they are prevented from becoming progressively colder each year by the:
   a. conduction of heat through the interior of the earth
   b. concentration of earth's magnetic field lines at the poles
   c. circulation of heat by the atmosphere and oceans
   d. the insulating properties of snow
   e. release of latent heat to the atmosphere when polar ice melts

ANSWER: c

34.   Suppose you drive to and from work on a street that runs east to west. On what day would you most likely have the sun shining directly in your eyes while driving to and from work?
        a. summer solstice
        b. winter solstice
        c. autumnal equinox
        d. during the summer months

ANSWER: c

35.   The term *normal* refers to weather data averaged over
        a. at least a day
        b. several months
        c. one year
        d. 30 years

ANSWER: d

36.   In July, at middle latitudes in the Northern Hemisphere, the day is ___ long and is ___ with each passing day.
        a. less than 12 hours, getting longer
        b. less than 12 hours, getting shorter
        c. more than 12 hours, getting longer
        d. more than 12 hours, getting shorter

ANSWER: d

37.   In meteorology, the word insolation refers to:
        a. a well-constructed, energy-efficient home
        b. the solar constant
        c. incoming solar radiation
        d. an increase in solar output

ANSWER: c

38.   During the afternoon the greatest temperature difference between the surface air and the air several meters above occurs on a:
        a. clear, calm afternoon
        b. clear, windy afternoon
        c. cloudy, calm afternoon
        d. cloudy, windy afternoon

ANSWER: a

39.   The greatest variation in daily temperature usually occurs:
        a. at the ground
        b. about 5 feet above the ground
        c. at the top of a high-rise apartment complex
        d. at the level where thermals stop rising

ANSWER: a

40.   In most areas the warmest time of the day about 5 feet above the ground occurs:
        a. around noon
        b. in the afternoon between 2 and 5 pm
        c. in the early evening after 6 pm
        d. just before the sun sets

ANSWER: b

41.   Everything else being equal, the lowest air temperature on a winter night will occur above a:
        a. surface covered with vegetation
        b. surface covered with snow
        c. bare surface
        d. surface covered with water

ANSWER: b

42.   The lowest temperature is usually observed:
        a. at the time of sunset
        b. near midnight
        c. several hours before sunrise
        d. around sunrise
        e. several hours after sunrise

ANSWER: d

43.   In clear weather the air next to the ground is usually ___ than the air above during the night, and
        _ than the air above during the day.
        a. colder, warmer
        b. colder, colder
        c. warmer, colder
        d. warmer, warmer

ANSWER: a

44. Suppose yesterday morning you noticed ice crystals (frost) on the grass, yet the minimum temperature reported in the newspaper was only 35° F. The <u>most</u> likely reason for this apparent discrepancy is that:
    a. temperature readings are taken in instrument shelters more than 5 feet above the ground
    b. the thermometer was in error
    c. the newspaper reported the wrong temperature
    d. the thermometer was read before the minimum temperature was reached for that day
    e. the thermometer was read incorrectly

ANSWER: a

45. Assuming that the night will remain clear, calm, and unsaturated, the predicted minimum temperature is 32° F. Suddenly the wind speed increases and remains gusty throughout the night. The minimum temperature will most likely be:
    a. about the same as predicted, but will occur earlier in the night
    b. higher than predicted due to the release of latent heat
    c. much lower than predicted due to radiational cooling
    d. higher than predicted due to mixing

ANSWER: d

46. At what time during a 24-hour day would a radiation temperature inversion best be developed?
    a. at sunset
    b. near sunrise
    c. toward the end of the morning
    d. between 2 and 5 pm when the air temperature reaches a maximum

ANSWER: b

47. The lag in daily temperature refers to the time lag between the:
    a. time of maximum solar radiation and the time of maximum temperature
    b. time of minimum temperature and the time of maximum solar radiation
    c. minimum and maximum temperature for a day
    d. minimum and maximum solar energy received at the surface for a given day
    e. sunrise and sunset

ANSWER: a

48. Ideal conditions for a strong radiation inversion:
    a. clear, calm, dry, winter night
    b. clear, calm, moist, summer night
    c. cloudy, calm, moist, winter night
    d. cloudy, windy, moist, summer night
    e. clear, windy, dry, summer night

ANSWER: a

49. Thermal belts are:
    a. pockets of warm air resting on a valley during the afternoon
    b. pockets of cold air resting on a valley floor at night
    c. warmer hillsides that are less likely to experience freezing conditions
    d. cold, below-freezing air found at the top of a mountain

ANSWER: c

50. The primary cause of a radiation inversion is:
    a. infrared radiation emitted by the earth's surface
    b. infrared radiation absorbed by the earth's surface
    c. solar radiation absorbed by the earth's surface
    d. solar radiation reflected by the earth's surface
    e. infrared radiation absorbed by the atmosphere and clouds

ANSWER: a

51. The deepest radiation inversion would be observed:
    a. at the equator any day of the year
    b. in polar regions in winter
    c. at the top of a high mountain in winter
    d. on a desert in winter
    e. in a deep valley during the summer

ANSWER: b

52. A radiation inversion is most commonly observed:
    a. when it is raining
    b. during the afternoon
    c. at sunset
    d. just above the ground
    e. in the upper atmosphere

ANSWER: d

53. On a clear, calm, night, the ground and air above cool mainly by this process:
    a. evaporation
    b. reflection
    c. convection
    d. conduction
    e. radiation

ANSWER: e

54. Which of the following can be used as a method of protecting an orchard from damaging low temperatures during a radiation inversion?
   a. orchard heaters
   b. wind machines
   c. irrigation (cover the area with water)
   d. all of the above

ANSWER: d

55. In a hilly region the best place to plant crops that are sensitive to low temperatures is:
   a. on the valley floor
   b. along the hillsides
   c. on the top of the highest hill
   d. in any dry location

ANSWER: b

56. Orchard heaters and wind machines are most useful in preventing damaging low temperatures from occurring next to the ground on:
   a. clear, windy nights
   b. cloudy, windy nights
   c. cloudy, snowy nights
   d. clear, calm nights
   e. rainy nights

ANSWER: d

57. Lines connecting points of equal temperature are called:
   a. isobars
   b. isotherms
   c. thermals
   d. thermographs

ANSWER: b

58. In summer, humid regions typically have ___ daily temperature ranges and ___ maximum temperatures than drier regions.
   a. smaller, lower
   b. smaller, higher
   c. larger, lower
   d. larger, higher

ANSWER: a

59. One would expect the lowest temperatures to be found next to the ground on a:
     a. clear, damp, windy night
     b. cloudy night
     c. clear, dry, calm night
     d. clear, dry, windy night
     e. rainy night

ANSWER: c

60. If tonight's temperature is going to drop into the middle 20s ($^{\circ}$F) and a fairly stiff wind is predicted, probably the best way to protect an orchard against a hard freeze is to (assume that cost is not a factor):
     a. use helicopters
     b. use wind machines
     c. sprinkle the crops with water
     d. put orchard heaters to work
     e. pray for clouds

ANSWER: c

61. Wind machines can prevent surface air temperatures from reaching extremely low readings by:
     a. blowing smoke over an orchard or field
     b. increasing the evaporation rate from fruits and vegetables
     c. mixing surface air with air directly above
     d. reducing the rate of cooling by evaporation
     e. increasing the likelihood of condensation on fruits and vegetables

ANSWER: c

62. An important reason for the large daily temperature range over deserts is:
     a. there is little water vapor in the air to absorb and re-radiate infrared radiation
     b. the light-colored sand radiates heat very rapidly at night
     c. dry air is a very poor heat conductor
     d. free convection cells are unable to form above the hot desert ground
     e. the ozone content of desert air is very low

ANSWER: a

63. Which of the following statements is(are) true?
    a. If you travel from Dallas, Texas to St. Paul, Minnesota in January, you are
       more likely to experience greater temperature variations than if you make the
       same trip in July
    b. Annual temperature ranges tend to be much greater near the ocean than in the
       middle of the continent
    c. If two cities have the same mean annual temperature, then their temperatures
       throughout the year are quite similar
    d. all of the above are true

ANSWER: a

64. Two objects A and B have the same mass but the specific heat of A is larger than B. If both
    objects absorb equal amounts of energy:
    a. A will become warmer than B
    b. B will become warmer than A
    c. both A and B will warm at the same rate
    d. A will get warmer, but B will get colder

ANSWER: b

65. The largest annual ranges of temperatures are found:
    a. at polar latitudes over land
    b. at polar latitudes over water
    c. at middle latitudes near large bodies of water
    d. at the Equator
    e. in the Northern Central Plains of the United States

ANSWER: a

66. Two objects have the same temperature. Object A feels colder to the touch than object B. This is
    probably because the two objects have different:
    a. thermal conductivities
    b. densities
    c. specific heats
    d. latent heats

ANSWER: a

67. This is used as a guide to planting and for determining the approximate date for harvesting crops:
    a. growing degree-days
    b. heating degree-days
    c. cooling degree-days
    d. mean annual temperature

ANSWER: a

64

68.   This is used as an index for fuel consumption:
        a. growing degree-days
        b. consumer price index
        c. heating degree-days
        d. mean annual temperature

ANSWER: c

69.   Which of the following is not a reason why water warms and cools much more slowly than land?
        a. solar energy penetrates more deeply into water
        b. heat energy is mixed in a deeper layer of water
        c. water has a higher heat capacity
        d. a portion of the solar energy that strikes water is used to evaporate it
        e. it takes more heat to raise the temperature of a given amount of soil 1° C than
           it does to raise the temperature of water 1° C.

ANSWER: e

70.   Over the earth as a whole, one would expect to observe the smallest variation in temperature from
        day to day and from month to month:
        a. at the North Pole
        b. in the center of a large land mass
        c. along the Pacific coast of North America
        d. high in the mountains in the middle of a continent
        e. on a small island near the equator

ANSWER: e

71.   How many heating degree-days would there be for a day with a maximum temperature of 30° F
        and a minimum temperature of 20° F? (Assume a base temperature of 65° F)
        a. 65
        b. 45
        c. 40
        d. 35
        e. 10

ANSWER: c

72.   How many cooling degree-days would there be for a day with a maximum temperature of 95° F
        and a minimum temperature of 65° F? (Assume a base temperature of 65° F)
        a. 30
        b. 25
        c. 20
        d. 15
        e. 0

ANSWER: d

73. Suppose peas are planted in Indiana on May 1. If the peas need 1200 growing degree-days before they can be picked, and if the mean temperature for each day during May and June is 70° F, in about how many days will the peas be ready to pick? (Assume a base temperature of 40° F)
    a. 30
    b. 40
    c. 70
    d. 120

ANSWER: b

74. In calm air the air temperature is -10° C, if the wind speed should increase to 30 knots (with no change in air temperature) the thermometer would indicate:
    a. a much higher temperature than -10° C
    b. a much lower temperature than -10° C
    c. a temperature of -10° C
    d. a temperature of -30° C

ANSWER: c

75. The air temperature is 45° F, the wind is blowing at 30 MPH, and the wind chill temperature is 15° F. These conditions would be equivalent to:
    a. a 15° F air temperature and 0 MPH winds
    b. a 30° F air temperature and 45 MPH winds
    c. a 30° F air temperature and 15 MPH winds
    d. a 15° F air temperature and 30 MPH winds

ANSWER: a

76. Hypothermia is most common in:
    a. hot, humid weather
    b. cold, wet weather
    c. hot, dry weather
    d. cold, dry weather

ANSWER: b

77. A thermometer that measures temperature and records it on a piece of chart paper:
    a. minimum thermometer
    b. thermistor
    c. thermograph
    d. maximum thermometer

ANSWER: c

78. The wind-chill factor:
    a. relates body heat loss with wind to an equivalent temperature with no wind
    b. indicates the temperature at which water freezes on exposed skin
    c. takes into account humidity and air temperature in expressing the current air temperature
    d. tells farmers when to protect crops from a freeze
    e. determines how low the air temperature will be on any given day

ANSWER: a

79. Which of the following is usually a liquid-in-glass thermometer?
    a. radiometer
    b. thermistor
    c. electrical resistance thermometer
    d. minimum thermometer
    e. thermograph

ANSWER: d

80. A thermometer with a small constriction just above the bulb is a(an):
    a. maximum thermometer
    b. minimum thermometer
    c. electrical thermometer
    d. thermocouple
    e. bimetallic thermometer

ANSWER: a

81. When would be the best time to reset a minimum thermometer?
    a. just after the time of minimum temperature
    b. just after the time of maximum temperature
    c. just before sunrise
    d. just before sunrise
    e. around noon

ANSWER: b

82. When a liquid thermometer is held in direct sunlight:
    a. it will accurately measure the air temperature
    b. it will measure a much higher temperature than that of the air
    c. it will measure a much lower temperature than that of the air
    d. it will measure the temperature of the sun rather than the air

ANSWER: b

83.   This instrument obtains air temperature by measuring emitted infrared energy:
        a. radiometer
        b. bimetallic thermometer
        c. electrical resistance thermometer
        d. thermistor
        e. thermograph

ANSWER: a

84.   The thermometer most likely to contain alcohol:
        a. bimetallic thermometer
        b. radiometer
        c. maximum thermometer
        d. thermograph
        e. minimum thermometer

ANSWER: e

85.   The thermometer that has a small dumbbell-shaped glass index marker in the tube is a:
        a. bimetallic thermometer
        b. maximum thermometer
        c. electrical thermometer
        d. thermocouple
        e. minimum thermometer

ANSWER: e

86.   An ideal shelter for housing a temperature-measurement instrument should be
        a. white
        b. black
        c. in the shade
        d. both (a) and (c)
        e. both (b) and (c)

ANSWER: d

Essay Exam Questions

1.   Describe the seasons that you would experience at two widely different locations on the earth. How might seasonal variations affect a region's culture and traditions?

2.   If you lived in the Arizona desert, would you want to own a black car? Why or why not? What might be a better color?

3.   An air temperature of 70° F feels quite comfortable. If you were in 70° F water, it would feel cold. Explain why.

4.   Explain why it is possible for the nighttime minimum to occur as much as 30 minutes after the sun has risen.

5.   What is the hottest (coldest) place in the United States (the world)? What factors cause these extreme conditions at these locations?

6.   Explain why it is possible to see frost on the ground or on the tops of parked automobiles even though the measured nighttime minimum temperature remains above 32° F.

7.   What meteorological conditions contribute to the formation of a strong radiation inversion? Why?

8.   How is it possible for the ground to become warmer than the air just above during the day and then turn colder than the air above during the night?

9.   What are the various methods used to protect sensitive crops from damaging low temperatures? Explain why each method works.

10.   A city which is located near a large body of water will generally have a milder climate than a city located at the same latitude in the center of a large mass. What factors account for this?

11.   Would a strong radiation inversion be more likely to form on a winter night or a summer night? Explain your answer.

12.   What types of temperature data might it be appropriate to include in a short description of a city or region's climate?

13.   Explain how the weather can affect people's emotions.

Chapter 4
Light, Color, and Atmospheric Optics

Summary

This chapter describes and explains a variety of atmospheric optical phenomena. The chapter begins with a brief review of the physical nature of light and explains our physiological perception of light and color. A first group of optical effects, which have the common characteristic that they are produced by scattering of light, are then discussed. Air molecules, for example, selectively scatter the shorter wavelengths of sunlight and give a clean sky its deep blue color. Larger aerosol particles scatter different wavelengths more equally, and can turn the sky milky white. White clouds, the blue color of distant mountains, and crepuscular rays are also examples of light scattering.

A second category of optical phenomena involves refraction and the dispersion of light. Mirages form when light is bent as it propagates through air layers with different densities. An inferior mirage can cause light from the sky to be bent so that it appears to be coming from the ground, and may make a road surface appear wet on a hot dry afternoon. Under unusual circumstances, refraction and dispersion can cause a green flash of light to appear as the sun sets below the horizon. Haloes and sundogs are relatively frequent events and occur when light passes through a high thin cloud layer composed of ice crystals. The formation of primary and secondary rainbows is discussed in some detail. In a primary rainbow, light rays are refracted as they enter a raindrop and are then reflected off the back inside surface of the raindrop. To see a rainbow, the sun must be low in the sky and at the observer's back.

The chapter concludes with a discussion of corona, iridescence, and the glory, all phenomena produced by the diffraction of light.

Key Terms

rods	green flash	critical angle
cones	mirage	internal reflection
reflected light	shimmer	primary rainbow
scattered light	inferior mirage	secondary bow
diffuse light	superior mirage	circumzenithal arc
Mie scattering	Fata Morgana	corona
selective scattering	halo	diffraction
Rayleigh scattering	22° halo	destructive interference
blue haze	46° halo	constructive interference
terpenes	tangent arc	iridescence
crepuscular rays	upper tangent arc	glory
transmitted light	lower tangent arc	brocken bow
refraction (of light)	dispersion (of light)	Heiligenschein
scintillation	sundogs (parhelia)	retroreflected (light)
twilight	sun pillars	
white night	rainbow	

Teaching Suggestions

Two books by C.F. Bohren are excellent sources of atmospheric optics demonstrations: *Clouds in a Glass of Beer*, John Wiley and Sons, New York, 1987; and *What Light Through Yonder Window Breaks*, John Wiley and Sons, New York, 1991. Another excellent source of optics and other atmospheric demonstrations is Z. Sorbjan's *Hands on Meteorology*, American Meteorological Society, Boston, 1996.

1.   Be sure that students understand that the color of an object is determined by the wavelength(s) of light reflected or scattered by the surface of the object (see Teaching Suggestion #7 in Chapter 2 of this manual). Place a red filter on the overhead projector (small sheets of gelatin filter material should be available at a local photographic supplies store). Place a green or blue object in the red light. The object will appear black. Place a white object in the red light and it will appear red.

2.   Many students will not appreciate the difference between reflection and scattering of light. Illuminate a smooth piece of aluminum foil with a beam of light. The beam will be reflected by the mirror-like surface. Next crumple the foil into a ball and then straighten and flatten it out. The light will now be scattered off in all directions by the wrinkled surface.

To demonstrate scattering, shine a beam of light (from a laser, a slide projector, or cover the top of an overhead projector with a piece of cardboard with a small circular aperture cut in the middle) through an aquarium or large beaker filled with water. If the water is clean and free of bubbles, the light beam of light will be invisible. Next, add a small amount of milk to the water. The beam of light will become clearly discernable. If the light source is white, the scattered light may also have a bluish tint. The fat globules in milk are sometimes small enough to selectively scatter short wavelengths. In this case, place a white screen at the far end of the aquarium to make the transmitted light visible to the class. The transmitted light will have a yellow or orange hue. If more milk is added, the beam of scattered light will broaden and become more diffuse; this is a demonstration of multiple scattering.

Shine a laser or beam of light across the front of a darkened classroom. Unless the air is unusually dusty, the beam will not be visible. Then clap two chalk board erasers together so that the chalk dust falls into the light beam. The beam will become visible. Or, hold a piece of dry ice above the beam so that the cloud that forms around the dry ice will descend into the light beam. This dense cloud is optically thick and scatters most of the incident light. Very little directly transmitted light will be visible on the opposite wall of the classroom.

3. Refraction can be demonstrated using a rectangular piece of thick (3/4 or 1 inch) plexiglas. Polish the edges of the plexiglas (suitable materials are often available from the plexiglas distributor). Press the plexiglas against a vertical screen. Using a laser or other narrow beam light source, cause light to shine down on the top edge of the plexiglas at an angle. With care the light source can be oriented so that the light beam just grazes the surface of the screen and its path will be visible. The light beam will bend noticeably as it enters the plexiglas and then again, in the opposite direction as it exits the plexiglas. Vary the angle of the incident beam of light.

A semi-circular piece of plexiglas can be used to show internal reflection as illustrated below. The incident light ray strikes the curved side of the plexiglas at a right angle and is not refracted. The angle at which the ray strikes the flat edge of the plexiglas can be varied to show internal reflection.

A circular piece of plexiglas can be cut on a lathe and used to illustrate the passage of a light ray through a raindrop.

4. Wait for a day when cirrostratus clouds are present in the sky, and take the class outdoors to look for a $22°$ halo and sundogs.

5. Demonstrate the dispersion of white light into its component colors using an equilateral prism. Cover the overhead with a large piece of cardboard which has a narrow slit cut in the middle. Hold the prism in the narrow beam of light between the projector and the screen. The spectrum will be projected onto the screen.

Student Projects

1. Many phenomena such as halos, sundogs and rainbows occur frequently enough that they can be observed and photographed by students.

2. Students might investigate additional phenomena not discussed in the chapter, such as polarization. Students could also investigate visibility in their locality and devise a means for estimating visibility (see Chapter 16 in Bohren's *Clouds in a Glass of Beer*).

3. Have students prepare a "recipe" for observing different kinds of optical phenomena. For example: near horizon and sun shining brightly, raining -- look for a rainbow opposite the sun.

 Blue Skies 4. Use the Sky Identification/Atmospheric Optics section of the BlueSkies cdrom to show photographs of several of the optical phenomena discussed in the textbook chapter. Have you ever seen any of these phenomena? To the best of your recollection, describe the weather conditions at the time you viewed these phenomena.

72

 Blue Skies 5.   Using the Sky Identification/Atmospheric Optics section of the BlueSkies cdrom, list the ice crystal shapes, and their fall orientations, responsible for five different optical phenomena caused by ice crystals.

Answers to Questions for Thought

1.   Before it rains the humidity may be high enough for water vapor to condense upon tiny particles and produce haze. Haze scatters all wavelengths of visible light; when viewed, haze appears white. The sky is a deeper blue after it rains because many of the haze particles are removed during the rainstorm. Also, if drier air follows behind the storm, the tiny particles are less likely to grow in size by condensation and, hence, they are less effective scatterers of all wavelengths of visible light. Air molecules and small particles are selective scatterers.

2.   Without an atmosphere to scatter light, there would be no twilight on the moon.

3.   The fog droplets effectively scatter light from the high beams back into the driver's eyes making it difficult to see.

4.   Red or orange.

5.   Planets are not hot enough to radiate visible light, the light we see from a planet is reflected light only. Stars are hot enough to emit visible light and the color of that light is dependent upon the star's temperature - the hotter the star, the more energy is emitted per unit area and the shorter the wavelength of maximum emission.

6.   A black sky at sunrise; the sun would always appear white (or slightly yellow-white).

7.   To see a rainbow, the sun must be at your back. The high sun at noon (especially in summer) makes this an almost impossible feat. Look at Fig. 4.27 and observe the angle at which sunlight enters and leaves a raindrop. A high sun puts the rainbow out of view for an observer on the ground.

8.   The blue haze is produced as particles much smaller than the wavelength of light selectively scatter only the shortest wavelengths of visible light (Rayleigh scattering). As the humidity increases, the particles grow larger by condensation and the larger particles are able to scatter all of the wavelengths of visible light about equally (Mie scattering).

9.   The sunlight that does reach the moon's surface has been refracted by the earth's atmosphere. Only the longest wavelengths are able to make it through the earth's thick atmosphere without being scattered; consequently, these waves make the moon's surface appear red when we view a lunar eclipse.

10.   Tiny smoke particles selectively scatter short waves and, thus, appear blue. In the mouth, moisture condenses on the smoke particles, they grow in size and become effective scatterers of all wavelengths of visible light. In the atmosphere these larger particles appear white.

11.   The *Novaya Zemlaya* effect occurs when the temperature changes dramatically above the ice-covered landscape. The changing temperature causes the refraction of light to change. Since the sun is

near the horizon, the bending of the sun's rays by the atmosphere can make it appear to rise above the horizon several days after it has set for the winter.

12.   Ultraviolet radiation is more intense because less of it has been scattered and absorbed by particles and air molecules.

13.   Moonlight is scattered making the night sky brighter and the stars less visible.

14.   During a moonlit night, the shorter wavelengths of visible light are scattered just as they are during the day. However, the intensity of the nighttime visible radiation is insufficient to be detected by the human eye.

Answers to Problems and Exercises

3.   (a) The milk has a blue cast (and the paper a red cast) because the tiny milk particles selectively scatter blue light and allow the longer (red) wavelengths to pass on through.
   (b) Milk particles are small - smaller than the wavelength of visible light.
   (c) Rayleigh scattering.
   (d) The atmosphere selectively scatters blue light turning the sky blue. Longer
      (red) wavelengths are transmitted, causing the sun to have a ruddy color.

Multiple Choice Exam Questions

1.   White light is ___ of electromagnetic radiation:
   a. a single long wavelength
   b. a single short wavelength
   c. a mixture of all visible wavelengths
   d. a mixture of all types

ANSWER: c

2.   Imagine that this piece of paper is illuminated with white light and appears red. You see red light because:
   a. the paper absorbs red and reflects other visible wavelengths
   b. the paper emits red light
   c. the paper reflects red and absorbs other visible wavelengths
   d. the paper disperses white light

ANSWER: c

3.   Beams of light that shine downward through breaks or holes in clouds are called:
   a. an inferior mirage
   b. crepuscular rays
   c. glorys
   d. corona

ANSWER: b

4.   On the average, as a cloud grows thicker (taller), which below does <u>not</u> occur?
        a. more sunlight is reflected from the cloud
        b. less sunlight is transmitted through the cloud
        c. less sunlight is absorbed by the cloud
        d. more light is scattered by the cloud

ANSWER: c

5.   Red sunsets, blue moons, and milky-white skies are <u>mainly</u> the result of:
        a. refraction
        b. dispersion
        c. reflection
        d. scattering
        e. diffraction

ANSWER: d

6.   Another name for diffuse light is:
        a. scattered light
        b. refracted light
        c. dispersion of light
        d. transmitted light

ANSWER: a

7.   The process that produces crepuscular rays in the atmosphere is:
        a. scintillation
        b. diffraction
        c. scattering
        d. dispersion
        e. refraction

ANSWER: c

8.   If the earth did not have an atmosphere, the sky would appear ___ during the day.
        a. white
        b. black
        c. red
        d. blue

ANSWER: b

9.   Stars are not visible during the day because:
        a. the scattered light coming from the sky is too bright to be able to see the
           weaker light from stars
        b. the earth is pointed away from the center of the galaxy
        c. the light from the stars is absorbed and scattered by the atmosphere and does
           not reach the ground
        d. all of the above

ANSWER: a

10.   The blue color of distant mountains is due primarily to:
        a. diffraction of light
        b. scattering of light
        c. refraction of light
        d. emission of light
        e. absorption of light

ANSWER: b

11.   Which of the following would be true if the earth did not have an atmosphere?
        a. there would be fewer hours of daylight
        b. the sky would always be black
        c. the stars would be visible in the sky during the day
        d. all of the above

ANSWER: d

12.   Air molecules selectively scatter visible light because:
        a. air molecules are smaller than the wavelength of visible light
        b. air molecules are much larger than the wavelength of visible light
        c. air molecules are the same size as the wavelength of visible light
        d. the electrons that orbit around the nucleus of atoms have a blue color

ANSWER: a

13.   The blue color of the sky is due to:
        a. selective scattering of visible light by air molecules
        b. the filtering effect of water vapor in the earth's atmosphere
        c. reflection of sunlight off the earth's oceans
        d. transmission of visible light through the ozone layer in the earth's stratosphere

ANSWER: a

14. On a foggy night, it is often difficult to see the road when the high beam lights are on because of ___ of light by the fog.
    a. absorption
    b. scattering
    c. transmission
    d. refraction
    e. diffraction

ANSWER: b

15. The sky is blue because air molecules selectively ___ blue light.
    a. scatter
    b. absorb
    c. diffract
    d. disperse
    e. emit

ANSWER: a

16. The nerve endings in the human eye responsible for perceiving colors are called
    a. pupils
    b. rods
    c. irises
    d. cones

ANSWER: d

17. What color would the sky be if air molecules selectively scattered only the longest wavelengths of visible light?
    a. white
    b. blue
    c. red
    d. black

ANSWER: c

18. Which of the following are capable of producing a red sunrise or sunset?
    a. small suspended salt particles
    b. volcanic ash
    c. small suspended dust particles
    d. all of the above

ANSWER: d

19. If the setting sun appears red, you may conclude that:
    a. the sun's surface temperature has cooled somewhat at the end of the day
    b. only the longest waves of visible light are striking your eye
    c. the next day's weather will be stormy
    d. you will not be able to see the moon that night

ANSWER: b

20. The "smoke" of the Great Smoky Mountains and the "blue" of the Blue Ridge Mountains are examples of
    a. sundogs
    b. glories
    c. halos
    d. haze

ANSWER: d

21. The sky will begin to turn milky white:
    a. when the concentration of ozone begins to reach dangerous levels
    b. when small particles such as dust and salt become suspended in the air
    c. when the relative humidity decreases below about ten percent
    d. on an oppressively hot day of the year

ANSWER: b

22. Oxygen and nitrogen
    a. scatter all wavelengths of light equally
    b. are selective scatterers
    c. are the only atmospheric gases that don't scatter any wavelengths of light

ANSWER: b

23. The bending of light that occurs when it enters and passes <u>through</u> a substance of different density is called:
    a. diffraction
    b. reflection
    c. refraction
    d. scattering

ANSWER: c

24.   When the sun is near the horizon the intensity of visible radiation reaching the earth's surface appears to be less than when the sun is directly overhead. Actually, the intensity of the visible radiation reaching the earth's surface is always the same.
    a. true
    b. false
    c. only at the equator

ANSWER: b

25.   Which of the following phenomena is not produced by refraction?
    a. halos
    b. crepuscular rays
    c. mirages
    d. sundogs
    e. none of the above

ANSWER: b

26.   Refraction of light by the atmosphere is responsible for:
    a. scintillation of starlight
    b. mirages
    c. causing the sun to appear to flatten-out on the horizon
    d. increasing the length of daylight
    e. all of the above

ANSWER: e

27.   Because of atmospheric refraction, a star seen near the earth's horizon is actually:
    a. slightly higher than it appears
    b. slightly lower than it appears
    c. much dimmer than it appears
    d. much further away than it appears

ANSWER: b

28.   When a beam of white light passes through a glass prism, it is separated into its component colors. This is called:
    a. diffraction
    b. dispersion
    c. selective scattering
    d. iridescence

ANSWER: b

29.   On a summer evening at middle latitudes, twilight adds about how much time to the length of daylight?
        a. 2 minutes
        b. 5 minutes
        c. 30 minutes
        d. 2 hours

ANSWER: c

30.   This phenomena can sometimes be seen near the upper rim of a setting or rising sun:
        a. sun pillar
        b. the glory
        c. a corona
        d. the green flash

ANSWER: d

31.   The green flash is largely an example of ___ of light by the earth's atmosphere.
        a. refraction
        b. reflection
        c. absorption
        d. diffraction

ANSWER: a

32.   An atmospheric phenomenon that causes objects to appear inverted is called:
        a. a superior mirage
        b. an inferior mirage
        c. scintillation
        d. dispersion

ANSWER: b

33.   A mirage is caused by:
        a. scattering of light by air molecules
        b. the bending of light by air of different densities
        c. a thin layer of moist air near the ground
        d. reflection of light from a hot surface

ANSWER: b

34. If the temperature was constant in the lowest 1000 meters of the atmosphere, conditions would be _____ for viewing a mirage.
   a. good
   b. excellent
   c. average
   d. poor

ANSWER: d

35. The Fata Morgana is actually a:
   a. mirage
   b. ice-crystal cloud
   c. rainbow
   d. sundog
   e. rainbow first seen in Morgan City, Utah

ANSWER: a

36. A wet-looking road surface on a clear, hot, dry day is an example of:
   a. a superior mirage
   b. scintillation
   c. diffraction
   d. condensation
   e. none of the above

ANSWER: e

37. Which of the following would you most likely observe over snow-covered ground in the winter?
   a. superior mirage
   b. sun pillars
   c. crespuscular rays
   d. shimmering

ANSWER: a

38. Which of the following are caused by the bending of light through ice crystals?
   a. rainbows and halos
   b. halos and the green flash
   c. halos and sundogs
   d. sundogs and sun pillars
   e. mirages and sundogs

ANSWER: c

39. A ring of light encircling the sun or moon could be either:
a. a rainbow or a halo
b. a halo or a sundog
c. a halo or a corona
d. a sundog or a crepuscular ray

ANSWER: c

40. An optical phenomenon that forms in a similar manner as the halo is the:
a. rainbow
b. corona
c. tangent arc
d. sun pillar
e. brocken bow

ANSWER: c

41. You would <u>most likely</u> see a tangent arc with a:
a. halo
b. sundog
c. rainbow
d. glory
e. corona

ANSWER: a

42. Halos are caused by:
a. refraction of light passing through raindrops
b. scattering of light by ice crystals
c. refraction of light passing through ice crystals
d. diffraction of light by cloud droplets
e. reflection of light by ice crystals

ANSWER: c

43. To see a sundog at sunrise, you should look toward the:
a. north
b. south
c. east
d. west

ANSWER: c

44.   To see a sundog, you should look about 22 degrees:
        a. to the right or left of the sun
        b. above the sun
        c. below the sun
        d. all of the above

ANSWER: a

45.   As light passes through ice crystals, ___ is bent the least and is, therefore, observed on the ___ of halos or sundogs.
        a. red, inside
        b. red, outside
        c. blue, inside
        d. blue, outside

ANSWER: a

46.   You would most likely see a halo or sundog with which of the following cloud types?
        a. altostratus
        b. cirrostratus
        c. nimbostratus
        d. cumulus

ANSWER: b

47.   Sun pillars are caused by ___ of light.
        a. dispersion
        b. diffraction
        c. scattering
        d. refraction
        e. reflection

ANSWER: e

48.   Sunlight reflecting off ice crystals produces this:
        a. crepuscular rays
        b. halos
        c. sun pillars
        d. sun dogs

ANSWER: c

49.   This can only be seen when the sun is to your back and it is raining in front of you:
       a. sundog
       b. halo
       c. rainbow
       d. sun pillar
       e. corona

ANSWER: c

50.   Secondary rainbows occur when:
       a. two internal reflections of light occur in raindrops
       b. light refracts through ice crystals
       c. a single internal reflection of light occurs in raindrops
       d. light refracts through a cloud of large raindrops
       e. the sun disappears behind a cloud and then reappears

ANSWER: a

51.   Clouds in the tropics tend to move from east to west. Consequently, which rhyme best describes
       a rainbow seen in the tropics?
       a. rainbow at the break of dawn, means, of course, the rain is gone
       b. rainbow at the break of day, means that the rain is on the way
       c. rainbow with a setting sun, means that sailors can have some fun
       d. rainbow in the morning, means that sailors should take warning

ANSWER: a

52.   Which of the following processes must occur in a raindrop to produce a rainbow?
       a. refraction, reflection, and dispersion of sunlight
       b. refraction, reflection, and scattering of sunlight
       c. reflection, scattering, and dispersion of sunlight
       d. transmission, reflection, and dispersion of sunlight
       e. refraction, transmission, and scattering of sunlight

ANSWER: a

53.   At sunset in the middle latitudes, look for a rainbow toward the:
       a. north
       b. south
       c. east
       d. west

ANSWER: c

54.   Which of the following is <u>not</u> involved in the formation of a rainbow?
        a. scattering
        b. refraction
        c. reflection
        d. dispersion
        e. diffraction

ANSWER: e

55.   Sun pillars are most commonly seen
        a. in very cold weather
        b. in very hot weather
        c. when it's raining
        d. in the tropics

ANSWER: a

56.   Which below is <u>not</u> true concerning a secondary rainbow?
        a. it is usually fainter than the primary rainbow
        b. it is seen above the primary rainbow in the sky
        c. the order of its colors is reversed compared to the primary rainbow
        d. the raindrops which produce the secondary rainbow are larger than the
           raindrops producing the primary bow

ANSWER: d

57.   It is _____ to produce an artificial rainbow with a garden hose.
        a. very easy
        b. almost impossible
        c. can't say - it depends on the type of hose.

ANSWER: a

58.   Which of the following is <u>true</u> about rainbows?
        a. the rainbow will be seen in the west when the sun is setting
        b. rainbows form when rays from the sun are scattered
        c. the brightest rainbows are seen around noon
        d. to see a rainbow at sunrise, you should look toward the west

ANSWER: d

59. Cloud iridescence is caused mainly by:
    a. refraction
    b. reflection
    c. diffraction
    d. dispersion
    e. scattering

ANSWER: c

60. Which atmospheric phenomenon below is produced by the diffraction of light around small water droplets?
    a. halo
    b. inferior mirage
    c. corona
    d. Heiligenshein
    e. sun pillar

ANSWER: c

61. Colored rings that appear around the shadow of an aircraft is called the:
    a. glory
    b. brocken bow
    c. Heiligenshein
    d. corona
    e. green flash

ANSWER: a

62. A faint ring of light that surrounds the shadow of an observer's head on a dew-covered lawn is called the:
    a. glory
    b. corona
    c. tangent arc
    d. brocken bow
    e. Heiligenshein

ANSWER: e

63. Which of the following is not caused by diffraction of light?
    a. glory
    b. cloud iridescence
    c. brocken bow
    d. corona
    e. tangent arc

ANSWER: e

64. Suppose you took a color photograph of clouds at night, with your camera adjusted so that the shutter stayed open long enough to allow enough the same amount of light that would enter the camera during a daytime photograph. On the resulting photograph, the clouds should look
a. dark
b. black and white
c. about the same as clouds look during a daytime photograph

ANSWER: c

Essay Exam Questions

1. Why does adding particles to the atmosphere affect visibility?

2. How might you demonstrate the phenomenon of light scattering using a glass tank filled with water, a flashlight, and some milk?

3. With a sketch, show why the setting sun will often appear red to an observer on the ground. Why does the sun appear white at noon, and red at sunrise and sunset?

4. Why aren't stars visible in the sky during the day?

5. Why do distant mountains appear blue?

6. Distinguish between the processes of reflection, refraction and scattering of light. Give an example of an atmospheric phenomena produced by each one. Which of these processes produces dispersion?

7. Illustrate with a sketch the refraction and dispersion of light as it passes through a glass prism.

8. Sketch the path that a ray of light follows as it passes through a raindrop and forms a primary rainbow. How is the path of a ray forming the secondary rainbow different?

9. Sketch a rainbow. Your drawing should show the correct positions of the primary and secondary bow and the proper order of colors in each. Where would the sun be in your picture?

10. What phenomena causes a road surface appear to be wet on a hot, clear, dry day? Why does the road appear to be wet? How would the road appear if it really were wet?

11. Using a diagram, if necessary, show why an inferior mirage produces an inverted image.

12. Would you expect to see a halo under clear or cloudy conditions? What does a halo tell you about upper atmospheric conditions?

13. What produces a sundog? Where and at what time of day should you look to try to see a sundog?

14. Describe the changes in appearance of a late-afternoon rainbow as the sun sinks toward the horizon.

Chapter 5
Atmospheric Moisture

Summary

The transformation of water from the gaseous to the liquid or solid state is an important source of energy in many meteorological processes and also makes weather phenomena visible to us. Some of the ways of measuring and expressing atmospheric water vapor concentrations are discussed in this chapter.

The chapter begins with descriptions of the different phases of water and the hydrologic cycle. The important concept of saturation is developed next. Saturation represents an effective upper limit to the amount of water vapor that may be found in air and is a function of air temperature. Several different parameters used to express the air's humidity including absolute humidity, specific humidity, mixing ratio, and water vapor pressure, are defined and explained. Because of recent trends in radio and television weather broadcasts, students are likely to have heard the terms "dew point" and "relative humidity". These concepts are compared, and students are shown that the dew point temperature provides a better absolute measure of the air's water vapor content than the relative humidity. The heat index, a practical measure of the effect that a combination of hot temperatures and high humidity have on our perception of temperature, is examined. Some of the techniques and instruments used to measure humidity are described at the end of the chapter. The chapter also includes interesting discussions of the average geographical variation of dew point temperature, and the effect of atmospheric moisture on baseballs in flight.

Key Terms

change of state
phase change
sublimation
deposition
evaporation
condensation
saturation
condensation nuclei
hydrologic cycle
transpiration
humidity
absolute humidity
specific humidity
mixing ratio
actual vapor pressure

Dalton's law of
  partial pressure
saturation vapor
  pressure
relative humidity
evaporative cooling
  systems
supersaturation
wet-bulb temperature
heat cramps
heat exhaustion
heat stroke
heat index (HI)
apparent temperature
atomic weight

dew point temperature
 (dew point)
frost point
psychrometer
sling psychrometer
aspiration psychrometer
dry-bulb temperature
wet-bulb depression
hygrometer
hair hygrometer
electrical hygrometer
infrared hygrometer
dew point hygrometer
dew cell

Teaching Suggestions

1.   The dew point temperature can be determined by filling a metal cup or container with warm tap water (at least $50°$ F in winter and $75°$ F in summer). Place a thermometer in the water and begin to slowly add ice while continuously stirring the mixture. Read the temperature when the first sign of condensation begins to appear on the outside of the container. This reading should be within about $3°$ F of the dew point temperature.

        The reading above could be compared with a simultaneous measurement made using an aspirated psychrometer and with the current dew point temperature announced on the local weather radio broadcast. Students could be asked to measure the dew point temperature and the relative humidity indoors and outdoors and asked to account for any similarities or differences.

2.   Saturation vapor pressure can be illustrated by filling a test tube with a few grams of solid iodine. The iodine will sublimate and fill the air in the test tube with iodine vapor. The iodine vapor has a purple-pink color which should just be visible at normal room temperature when viewed against a white background.

        The effect of temperature on the saturation vapor pressure can be demonstrated by immersing the bottom of a second test tube in hot water. The gas in the test tube will have a noticeably darker color indicating a higher vapor concentration. If this warm test tube is cooled, iodine vapor will be deposited on the sides of the test tube.   With some care two test tubes (warm and cool) may be passed around a small class for examination. The iodine vapor is also visible when the test tubes are placed on an overhead projector.

3.   Measure relative humidity with a sling psychrometer and a psychometric table. Discuss the difference between dew point and wet bulb temperatures. Why are there different psychometric tables for different atmospheric pressures?

Student Projects

1.   Have students plot daily high and low temperatures together with average dew point temperature and an estimate of cloud cover for a period of a week or two. Were the coldest nighttime temperatures observed with clear or cloudy skies, dry or humid conditions? Have students plot the daily temperature range against one of the moisture variables. How much variation is there in the dew point temperature from day to day? Can students explain sudden changes in the daily average dew point? Is there any correlation between apparent visibility and dew point temperature?

2.   Add the relative humidity (corresponding to the high and low temperatures) to the above graphs. Which moisture variable, dew point or RH, gives a better indication of how humid the air feels?

Blue Skies 3.   Use the Moisture and Stability/Moisture Graph activity on the BlueSkies cdrom to demonstrate the relationship between temperature, vapor pressure and relative humidity during heating, cooling, evaporation and condensation. Describe these relationships in your own words.

Blue Skies 4.   Use the Moisture and Stability/Adiabatic exercise on the BlueSkies cdrom to answer the following questions.
(a) With a temperature of 10.0°C and a dew point temperature of 0.0°C on the upwind side of the Sierra Nevada mountain range, what will be the altitude of cloud formation on the upwind side, and the ground-level temperature on the downwind side, of air crossing the mountain range? (Answer: 1300 m; 17.0°C)
(b) Same as above, but use a temperature of 10°C and a dew point temperature of 5°C. (Answer: 700 m; 19.5°C).

Answers to Questions for Thought

1.   The water should evaporate from the glass more quickly on the windy, warm, dry summer day.

2.   The clothes "dry" by sublimation - ice to vapor transformation at subfreezing temperatures.

3.   (a) Absolute humidity decreases because the mass of water vapor remains constant as the volume of air increases.
       (b) The relative humidity increases as the parcel cools because the air approaches the dew-point temperature and the parcel approaches saturation.
       (c) Actual vapor pressure decreases slightly due to the reduced pressure inside the rising, expanding parcel.
       (d) Saturation vapor pressure decreases as the air cools because of the lower air temperature. (Each of these examples assumes that the air remains unsaturated)

4.   According to Fig. 5.12, the smallest difference in the average water vapor pressure between July and January would be in Nevada. Overall, the difference in vapor pressure between these two months is less in the western one-third of the United States than in the eastern two-thirds of the United States.

5.   Sure. Nighttime temperatures can get quite low in the desert, causing the relative humidity to rise above 90%.

6.   The dew point temperature is a measure of the air's actual water vapor content. Water evaporating into the air increases the air's water vapor content and the dew point rises.

7.   The cooler the air (morning) the closer the air is to being saturated with water vapor and the higher is its relative humidity. The warmer the air (afternoon) the farther it is from being saturated and the lower its relative humidity.

8.   The relative humidity indoors would be higher than before. The amount of water vapor in the air did not change. However, the cooler air indoors is now closer to being saturated.

9.   The air is close to being saturated. The air temperature and dew point are close together.

10.   In the dry air, the wick on the wet-bulb thermometer has dried out and the temperature reading is no longer the wet-bulb temperature.

11.   The climate is quite dry in much of Arizona, Nevada and California, thus the air is usually far from saturated with respect to liquid water. In these conditions evaporative coolers are quite effective in lowering the temperature by removing the latent heat of evaporation from the air. In Florida, Georgia and Indiana, the climate is much more humid, thus evaporative coolers are not nearly as efficient.

12.   Boil water in the pot and measure the boiling-point temperature. Use Fig. 1 in the Focus section "Vapor Pressure and Boiling" in the text to determine your pressure level. Relate this pressure to elevation (for altitudes near sea level, use a vertical pressure change of 10 mb/100 m).

Answers to Problems and Exercises

1.   RH = 100% x (3.7 mb / 25 mb) = 15%

2.   (a) St. Louis, about 18 $^{\circ}$C (64 $^{\circ}$F)
            New Orleans, about 22 $^{\circ}$C (72 $^{\circ}$F)
            Los Angeles, about 13 $^{\circ}$C (55 $^{\circ}$F)
        (b) e_s = 48 mb when air temperature is 32 $^{\circ}$C
            St. Louis, RH = 100% x (21 mb)/(48 mb) = 44%
            New Orleans, RH = 100% x (27 mb)/(48 mb) = 56%
            Los Angeles, RH = 100% x (15 mb)/(48 mb) = 31%

3.   (a) wet bulb depression = 5 $^{\circ}$C
        (b) dew point temperature = 23 $^{\circ}$C
        (c) relative humidity = 67%

4.   (a) e_s = 56.2 mb, e = 25 mb, RH = 100% x (25 mb)/(56.2 mb) = 44%
        (b) e_s = 56 mb, e = 25 mb, RH = 100% x (25 mb)/(56 mb) = 44%
        (c) A dry bulb of 35 $^{\circ}$C and a dew point of 21 $^{\circ}$C corresponds to a wet-bulb depression of 10 $^{\circ}$C. An air temperature of 35 $^{\circ}$C with a wet-bulb depression of 10° yields a RH of 44%.

5.   (a) about 4 °C (40 °F)
        (b) about 9 mb

6.   From Fig. 1 in the Focus section "Vapor Pressure and Boiling" about 92.5 °C.

7.   About 93 °F.

8.   (a) Seattle: 52 °F; New York City: 66°F
        (b) From Figure 5.10, $e_s(52\,°F) = 13$ mb and $e_s(66\,°F) = 22$mb. Thus the abundance of water vapor in New York City is (22-13)/13 = 69%.

Multiple Choice Exam Questions

1.   If a glass of water were surrounded by saturated air:
        a. the level of the water in the glass would slowly decrease
        b. the water's temperature would slowly increase
        c. the level of the water in the glass would not change
        d. the water's temperature would slowly decrease

ANSWER: c

2.   When the air is saturated, which of the following statements is not correct?
        a. the air temperature equals the wet-bulb temperature
        b. the relative humidity is 100%
        c. the air temperature equals the dew point temperature
        d. an increase in temperature will cause condensation to occur.
        e. the wet bulb temperature equals the dew point temperature

ANSWER: d

3.   As the air temperature increases, the air's capacity for water vapor:
        a. increases
        b. decreases
        c. remains constant
        d. is unrelated to air temperature and can either increase or decrease

ANSWER: a

4.   If all the water vapor in the atmosphere were to condense and fall to the ground, the globe would be covered with about ___ of water.
        a. 1 millimeter
        b. 1 inch
        c. 1 foot
        d. 1 meter

ANSWER: b

5.   The total mass of water vapor stored in the atmosphere at any moment is about ___ of the world's supply of precipitation.
     a. 1 day
     b. 1 week
     c. 1 month
     d. 1 year

ANSWER: b

6.   If the amount of water vapor in the atmosphere remains constant, the humid
     a. cannot change
     b. must change
     c. might change, depending on which measure of humidity is used
     d. will only change if we're using absolute humidity as the measure of humidity

ANSWER: d

7.   The density of water vapor in a given parcel of air is expressed by the:
     a. absolute humidity
     b. relative humidity
     c. mixing ratio
     d. specific humidity
     e. saturation vapor pressure

ANSWER: a

8.   Which of the following will increase in a rising parcel of air?
     a. saturation vapor pressure
     b. relative humidity
     c. mixing ratio
     d. air temperature
     e. none of the above

ANSWER: b

9.   Which of the following will decrease in a rising parcel of air?
     a. relative humidity
     b. absolute humidity
     c. specific humidity
     d. all of the above

ANSWER: b

10.   The ratio of the mass of water vapor in a given volume (parcel) of air to the mass of the remaining dry air describes the:
       a. absolute humidity
       b. mixing ratio
       c. relative humidity
       d. dew point

ANSWER: b

11.   When the air temperature increases, the saturation vapor pressure will:
       a. increase
       b. decrease
       c. remain the same
       d. vary over an increasingly broad range of values

ANSWER: a

12.   The maximum pressure that water vapor molecules would exert if the air were saturated is called the:
       a. absolute humidity
       b. boiling point
       c. mixing ratio
       d. none of the above

ANSWER: d

13.   If water vapor comprises 3.5% of an air parcel whose total pressure is 1000 mb, the water vapor pressure would be:
       a. 1035 mb
       b. 35 mb
       c. 350 mb
       d. 965 mb

ANSWER: b

14.   A high water vapor pressure indicates:
       a. a relatively large number of water vapor molecules in the air
       b. a relatively small number of water vapor molecules in the air
       c. a relatively high rate of evaporation
       d. an abundant supply of condensation nuclei in the air

ANSWER: a

15.   If the air temperature increased, with no addition or removal of water vapor, the actual vapor pressure would:
        a. increase
        b. decrease
        c. stay the same
        d. become greater than the saturation vapor pressure

ANSWER: c

16.   When the air temperature is below freezing, the saturation vapor pressure over water is ___.
        a. equal to zero
        b. less than the saturation vapor pressure over ice
        c. greater than the saturation vapor pressure over ice
        d. equal to the saturation vapor pressure over ice

ANSWER: c

17.   Ignoring the contributions of wind, a baseball would be expected to travel farther in warm, humid air because
        a. water vapor helps to lift the baseball even higher
        b. water vapor sticks to baseballs
        c. water vapor is less dense than dry air

ANSWER: c

18.   A baseball hit several hundred feet on a hot, humid day will travel about _____ it would have traveled on a hot, dry day.
        a. 1%
        b. 10%
        c. 50%
        d. 100%

ANSWER: a

19.   The Gulf Coast states are more humid in summer than the coastal areas of Southern California mainly because of the:
        a. higher air temperature in the Gulf States
        b. lower air temperature in Southern California
        c. higher water temperature in the Gulf of Mexico
        d. low relative humidity of the air over the Pacific Ocean

ANSWER: c

20. If very cold air is brought indoors and warmed with no change in its moisture content, the saturation vapor pressure of this air will ___ and the relative humidity of this air will ___.
    a. increase, increase
    b. decrease, decrease
    c. increase, decrease
    d. decrease, increase

ANSWER: c

21. Evaporative coolers are primarily used in climates where the summers are:
    a. hot and humid
    b. hot and dry
    c. cold and humid
    d. cold and dry

ANSWER: b

22. Which of the following will increase the relative humidity in a home during the winter?
    a. increasing the thermostat setting
    b. lowering the temperature of the air inside the home
    c. sealing the house against drafts
    d. taking a shower and letting the air circulate through the home

ANSWER: d

Questions 23 through 26 refer to the temperature and dew point data in the following cities:

City	Air Temperature ($^\circ$F)	Dew Point ($^\circ$F)
City A	95	76
City B	10	10
City C	30	21
City D	50	42

23. Which city has the highest relative humidity?
    a. City A
    b. City B
    c. City C
    d. City D

ANSWER: b

24. Which city has the least amount of water vapor in the air?
    a. City A
    b. City B
    c. City C
    d. City D

ANSWER: b

25.   Which city has the <u>greatest</u> amount of water vapor in the air?
       a. City A
       b. City B
       c. City C
       d. City D

ANSWER: a

26.   Which city has the <u>highest</u> saturation vapor pressure?
       a. City A
       b. City B
       c. City C
       d. City D

ANSWER: a

27.   The main reason why vegetables take longer to cook in boiling water at high altitudes is **because**:
       a. water boils at a higher temperature with higher altitude
       b. the temperature of the boiling water decreases with increasing altitude
       c. there is less oxygen in the air at high altitude
       d. saturation vapor pressure decreases with increasing altitude

ANSWER: b

28.   The temperature at which water boils depends mainly on:
       a. air temperature
       b. relative humidity
       c. air pressure
       d. air density
       e. the specific heat of air

ANSWER: c

29.   The percentage of water vapor present in the air compared to that required for saturation is the:
       a. mixing ratio
       b. absolute humidity
       c. dew point
       d. relative humidity
       e. specific humidity

ANSWER: d

30.   Suppose it is snowing outside and the air is saturated. The air temperature and dew point are both 15 °F, and the actual vapor pressure is 3 mb. If this air is brought indoors and warmed to 75 °F, what would the relative humidity of this air be, assuming that its moisture content does not change? (The saturation vapor pressure at 75 °F is 30 mb).
   a.   5 percent
   b. 10 percent
   c. 30 percent
   d. 50 percent
   e. 100 percent

ANSWER: b

31.   At what time of day is the relative humidity normally at a minimum?
   a. when the air temperature is highest
   b. just before sunrise
   c. about midnight
   d. when the air temperature is lowest

ANSWER: a

32.   The time of day when the relative humidity reaches a maximum value is usually:
   a. at the time when the air temperature is highest
   b. in the middle of the afternoon
   c. at the time when the air temperature is lowest
   d. just before sunrise
   e. about midnight

ANSWER: c

33.   The dew point temperature is a measure of the total amount of water vapor in the air.
   a. true
   b. false

ANSWER: a

34.   As the air temperature increases, with no addition of water vapor to the air, the relative humidity will:
   a. remain the same
   b. increase
   c. decrease
   d. increase until it becomes equal to the dew point temperature

ANSWER: c

35. The relative humidity is often near 100% in the polar regions.
    a. true
    b. false

ANSWER: a

36. With which set of conditions below would you expect wet laundry hanging outdoors on a clothesline to dry most quickly?

	Air Temperature (°F)	Relative Humidity	Wind Speed
a.	60	75%	20 MPH
b.	40	75%	20
c.	60	50%	20
d.	40	50%	10
e.	60	75%	10

ANSWER: c

37. If the air temperature remains constant, evaporating water into the air will ___ the dew point and ___ the relative humidity.
    a. increase, increase
    b. increase, decrease
    c. decrease, increase
    d. decrease, decrease

ANSWER: a

38. Suppose the dew point of cold outside air is the same as the dew point of the air indoors. If the door is opened and cold air replaces some of the warm air, then the new relative humidity indoors would be:
    a. lower than before
    b. higher than before
    c. the same as before
    d. impossible to tell from the information given

ANSWER: b

39. If the air temperature in a room is 70° F, the saturation vapor pressure is 25 mb, the dew point temperature is 45° F, and the actual vapor pressure is 10 mb, then the relative humidity must be near ___ percent.
    a. 15
    b. 20
    c. 35
    d. 40

ANSWER: d

40.   Suppose saturated polar air has an air temperature and dew point of -10° C, and unsaturated desert air has an air temperature of 35° C and a dew point of 10° C. The desert air contains ___ water vapor and has a ___ relative humidity than the polar air.
       a. more, lower
       b. more, higher
       c. less, lower
       d. less, higher

ANSWER: a

41.   A sling psychrometer *directly measures* the relative humidity.
       a. true
       b. false

ANSWER: b

42.   As the difference between the air temperature and the dew point increases, the relative humidity:
       a. increases
       b. decreases
       c. remains constant at a value less than 100%
       d. remains constant and equal to 100%

ANSWER: b

43.   The temperature to which air must be cooled in order to become saturated is the:
       a. minimum temperature
       b. dew point temperature
       c. wet-bulb temperature
       d. freezing point

ANSWER: b

44.   As the air temperature increases, with no addition of water vapor to the air, the dew point will:
       a. remain the same
       b. increase
       c. decrease
       d. increase and become equal to the air temperature

ANSWER: a

45. In a blinding snowstorm in Vermont the air temperature and dew-point temperature are both 30° F. Meanwhile, under clear skies in Arizona, the air temperature is 85° F and the dew point temperature is 38° F. From this information you could conclude:
   a. there is more water vapor in the air in Arizona
   b. there is more water vapor in the Vermont snowstorm
   c. the same amount of water vapor is found in the air in Vermont and Arizona
   d. Vermont and Arizona are both located next to the ocean

ANSWER: a

46. Which of the following is the best indicator of the actual amount of water vapor in the air?
   a. air temperature
   b. saturation vapor pressure
   c. relative humidity
   d. dew point temperature

ANSWER: d

47. At 40° F, the atmosphere is saturated with water vapor. If the air temperature increases to 60° F, with no addition or removal of water vapor, one may conclude that the dew point is about:
   a. 20° F
   b. 40° F
   c. 60° F
   d. 100° F

ANSWER: b

48. The lowest temperature that can be attained by evaporating water into the air is known as the:
   a. heat index
   b. minimum temperature
   c. wet-bulb temperature
   d. frost point
   e. wind chill temperature

ANSWER: c

49. This instrument uses wet-bulb and dry-bulb temperature to obtain relative humidity:
   a. infrared hygrometer
   b. sling psychrometer
   c. hair hygrometer
   d. electrical hygrometer

ANSWER: b

50.   Which of the following statements is <u>not</u> correct?
       a. The length of human hair changes as the relative humidity changes.
       b. During the winter, low relative humidities can irritate the mucus membranes
          in the nose and throat.
       c. The relative humidity is a measure of the air's actual water vapor content.
       d. A change in the air temperature can change the relative humidity.

ANSWER: c

51.   The instrument that measures humidity by measuring the amount of radiant energy absorbed by
       water vapor is the:
       a. electrical hygrometer
       b. infrared hygrometer
       c. sling psychrometer
       d. hair hygrometer
       e. dew cell

ANSWER: b

52.   Nighttime temperatures rarely drop below the dew point temperature because
       a. the dew will absorb all the heat
       b. saturation vapor pressures always increase at night
       c. at saturation, latent heat of condensation is released into the air
       d. both (b) and (c)

ANSWER: c

Essay Exam Questions

1.   With the aid of a figure, illustrate and describe the circulation of water in the atmosphere (the
      hydrologic cycle).

2.   Bespectacled people know that when entering a warm building after being outdoors in
      subfreezing temperatures, condensation occurs on the lenses of their glasses. Why doesn't
      condensation occur on the lenses when they go outside after a lengthy stay inside the warm
      building?

3.   What is meant by the terms water vapor saturation and saturation vapor pressure? Why does the
      saturation vapor pressure increase with increasing air temperature?

4.   Explain why, at temperatures below freezing, the saturation vapor pressure over water is greater
      than the saturation vapor pressure over ice.

5.   In terms of the air temperature and water vapor content, explain how the relative humidity
      normally changes during the course of a 24-hour day.

6. In order to reduce evaporation and conserve water, do you think it would be better to water a lawn in the early morning or early evening?

7. Would lowering the temperature in your home during the winter cause the relative humidity to increase or decrease? Why? What physiological effects might you experience during humid and dry conditions?

8. Explain why the dew point temperature provides a better indication of the actual amount of water vapor in the air than the relative humidity.

9. Describe how a sling psychrometer can be used to determine the relative humidity or dew point.

10. Would hazy conditions generally indicate dry or humid conditions?

11. An enterprising travel agent trying to sell you a summer vacation package to Phoenix, Arizona, claims that the summertime relative humidity is above 90%. Can this be true? Explain.

Chapter 6
Condensation: Dew, Fog, and Clouds

Summary

The formation of dew, frost, and various types of fog and clouds are discussed in this chapter. Dew or frost form when moist air next to the ground is cooled to or below the dew point temperature and water vapor condenses onto objects on or near the ground. Meteorological conditions that favor the formation of dew or frost are examined.

The role that condensation nuclei play in cloud formation is considered next. We see that the formation of fog is a progressive process that starts at relative humidities less than 100% when water vapor begins to condense onto hygroscopic condensation nuclei in the air. Cloud droplets scatter light and a cloud layer that forms near the ground is officially designated haze or fog depending on the reduction in visibility that it causes. Fog can be produced under a variety of situations either by cooling moist air to saturation or by evaporating or mixing water vapor into the air. Factors that may make a region prone to fog formation are examined; techniques used to disperse fog are discussed in a focus section.

In the next section, students are shown that a seemingly infinite variety of cloud forms can be classified into ten basic types according to their appearance and the altitude at which they form. An interesting focus section describes how cloud ceiling can be measured. Photographic illustrations of the basic cloud types and several unusual cloud forms are given. The chapter ends with a discussion of weather satellites and the kinds of information that may be derived from them. Many color satellite images are included as illustrations.

Key Terms

dew point
dew
frozen dew
frost point
deposition
hoar frost
white frost
frost
freeze
black frost
condensation nuclei
Aitken nuclei
cloud condensation nuclei
hygroscopic nuclei
hydrophobic nuclei
haze
dry haze
wet haze
fog
acid fog
radiation (ground) fog
valley fog
high inversion fog
advection fog
advection-radiation fog
ice fog
upslope fog
evaporation fog
evaporation (mixing) fog
steam fog
steam devils

arctic sea smoke
frontal fog
winter chilling
fog lamps
cold fog
supercooled fog
warm fog
turboclair
stratus
cumulus
cirrus
nimbus
cirrus clouds
mare's tails
cirrocumulus clouds
mackerel sky
cirrostratus clouds
altocumulus clouds
castellanus
altostratus clouds
watery sun
nimbostratus clouds
stratus fractus
scud
stratocumulus clouds
stratus clouds
cumulus clouds
cumulus humilis
cumulus fractus
cumulus congestus
towering cumulus

cumulonimbus clouds
cumulonimbus incus
lenticular clouds
banner clouds
cap cloud
pileus clouds
mammatus clouds
contrails
aerodynamic contrail
nacreous clouds
mother-of-pearl clouds
noctilucent clouds
scattered (cloud cover)
broken (cloud cover)
overcast (cloud cover)
ceiling
ceiling balloons
ceilometer
geostationary satellites
geosynchronous satellites
polar orbiting satellites
TIROS I
GOES (Geostationary
  Operational Environmental
  Satellite)
infrared cloud pictures
computer enhancement
METEOSAT
LandSat

Teaching Suggestions

1.   A dramatic demonstration of cloud formation in a bottle has been described by C.F. Bohren (*Clouds in a Glass of Beer*, pp. 8-14, John Wiley and Sons, New York, 1987). Place a small amount of water in a large bottle or flask. Close the top of the bottle with a rubber stopper. Connect a pump (an ordinary bicycle pump works well) to the bottle with a tube through a hole in the stopper. Pump air into the bottle. If the stopper has not been inserted too firmly, it will suddenly pop as air is being pumped into the bottle and will allow the air to expand outward and cool. A faint cloud should be visible in the bottle.

        Repeat the experiment, but this time add hygroscopic smoke particles by dropping a lighted match into the bottle prior to pressurization. This time a cloud will be much thicker and more clearly visible. The visibility is enhanced if a light source is placed behind the bottle so that the students are in a position to see light scattered in the forward direction by the cloud droplets. The cloud will disappear when the air in the bottle is pressurized. The compression warms the air and mixes in drier air from outside the bottle.

This demonstration can be used to explain the formation of haze, fog, and clouds, as well as the role that condensation nuclei play in their development. The article by Bohren also includes a good demonstration of cloud droplet formation on grains of salt.

2.   Dry ice offers a convenient way of producing thick clouds. Ask the students why a thicker cloud is formed when one blows on the piece of dry ice. Students should also understand why the cloud sinks.

3.   Photographic slides greatly enhance a discussion of clouds and cloud type identification. Use a portion of the cloud slide collection (available to the instructor as a supplement to the textbook) to illustrate the common cloud types and their characteristics. Then show additional cloud slides and ask the students to identify the cloud type. If possible, include photographs of cloud patterns from the local area. This will give the students some experience identifying clouds they are likely see at some point during the course. In some areas, students should be made aware of local topography that can be used to judge low cloud base heights. In many cities it is possible to have slide photographs processed overnight. In this way, interesting weather features or events can be discussed in a timely manner in class.

Blue Skies 4.   Using the Sky Identification/Name that Cloud section of the BlueSkies cdrom, invite students to participate in a "Name that Cloud" competition during class.

Student Projects

1.   During certain times of the year, students can observe the formation of dew, frost or fog. Have the students record weather data on these days as well as days on which dew, frost and fog did not form.

Is frost ever observed when the measured overnight minimum temperature remains above freezing? Is the layer of dew or frost thicker on some objects than other objects? Does the color or composition of the object have any effect on the formation of dew or frost? Is the formation of dew or frost thicker at ground level or on objects above ground level? A book by C.F. Bohren (ref: *What Light Through Yonder Window Breaks*, pp. 97-111, John Wiley and Sons, New York, 1991) contains an excellent discussion of how environmental factors affect the formation of dew and frost. Many of his experiments and observations could serve as student projects.

2.   Students should be encouraged to observe and try to identify different types of clouds. Students could observe and record the sequence of different types of clouds that are associated with the approach and passage of a low pressure center or front. If they have sufficient means, students should be encouraged to photograph interesting cloud patterns and weather phenomena.

3.   Using local landmarks, make a daily estimate of visibility. Determine whether visibility values are correlated with meteorological variables such as relative humidity.

Blue Skies 4.   Use the Weather Forecasting/Forecasting section of the BlueSkies cdrom to find the current surface air temperature and dew-point temperature in your area Is fog likely to form today? Why or why not?

Blue Skies 5.   Use the Atmospheric Basics/Layers of the Atmosphere section of the BlueSkies cdrom to examine a sounding (vertical profile of temperature, dew point temperature, etc.) for a location near you. Are clouds likely to form? At what altitude(s)? What type of clouds are they likely to be?

Answers to Questions for Thought

1.   A cloud-free nighttime sky and a clearly visible moon indicate that the air is dry and that rapid radiational cooling will take place near the ground.

2.   Icebergs chill the air in contact with them to below the dew-point temperature.

3.   The sunglasses cool to the air temperature inside the motel room. Once you step outside, the temperature of the sunglasses is below the dew-point temperature of the humid outside air. The glasses cool the air in contact with them to the dew-point temperature, and moisture condenses onto them.

4.   The windshield is quite cold. The car apparently travels into warmer air with a higher dew-point. The outside windshield cools the warmer air in contact with it to its dew-point temperature which is still below freezing and frost forms.

5.   Really clean atmospheres would have no condensation nuclei so that clouds would not be able to form. Really dirty atmospheres are unhealthy.

6.   In polluted air, condensation begins on hygroscopic particles when the relative humidity is less than 100 percent. Condensation removes water vapor from the air which lowers the dew-point temperature. This increases the spread between air temperature and dew-point temperature, and keeps the relative humidity at less than 100 percent.

7.   Tropical waters are usually too warm to cool the air to its dew-point temperature.

8.   The cold snow surface cooled relatively warm, moist air (moving in from the south) to its dew-point temperature.

9.   Fog droplets are so tiny that they settle very slowly. As long as the air remains saturated, fog is maintained as water vapor condenses onto nuclei forming new fog droplets.

10.   Steam fog forms when cool air moves over a large body of warmer water. Due to the thermal properties of water, the water remains relatively warm during the autumn months. On the other hand, advection fog forms when warm, moist air moves over a colder surface. In spring the water of a large lake will warm more slowly and will remain colder than the adjacent land.

11.   Latent heat of condensation is released as fog droplets form. Also, fog absorbs infrared energy radiated from the surface and re-radiates this energy back toward the surface.

12.   The dew point decreased because water vapor was being removed from the air and transformed into liquid fog droplets. The dew point is a measure of the amount of water vapor in the air.

13.   You see your breath because the warm, moist air from your mouth mixes with cooler air in the atmosphere. The mixture is saturated and condensation occurs with the exhaled breath. The air temperature does <u>not</u> have to be below freezing for this occur.

14.   The air temperatures aloft are much lower above polar latitudes than above subtropical latitudes. Altocumulus clouds usually contain liquid water. Over polar latitudes at 6400 m (21,000 feet), temperatures are generally low enough to freeze all liquid water. Therefore, the very cold air will produce a thin, ice crystal (cirriform) cloud. At 21,000 feet above the tropics and subtropics temperatures are not low enough to freeze all liquid water; consequently, thicker middle clouds may form at that altitude.

15.   If the rain is light or moderate and steady, the cloud is most likely nimbostratus. If the rain falls in heavy showers, and lightning, thunder, or hail accompany it, then the cloud is probably cumulonimbus. (Careful: cumulus congestus may also produce heavy showers)

16.   Cumulus fractus (or stratus fractus). Rain falling from the nimbostratus cloud evaporates, mixes, and saturates the air beneath it. If slight vertical motions exist, a cloud often forms.

Answers to Problems and Exercises

1.   (a) Greatest likelihood of visible frost on Morning #4 because the dew point (frost point) will be reached at below-freezing temperatures.
        (b) Frozen dew is most likely on Morning #3 because the air temperature will reach the dew point at 1 °C, dew will form and then it will freeze when the air cools to 0 °C.
        (c) A black frost is most likely on Morning #2 because the air temperature is expected to drop below freezing but not reach the dew point.
        (d) Dew would be most likely to be observed on the ground on Morning #5 because the air temperature is expected to reach the dew point, but it is not expected to drop to or below freezing.

2.   The ceiling is about 600 m (about 2000 feet) above ground level.

Multiple Choice Exam Questions

1. Dew is most likely to form on:
   a. clear, calm nights
   b. cloudy, calm nights
   c. clear, windy nights
   d. cloudy, windy nights
   e. rainy nights

ANSWER: a

2. The name given to a liquid drop of dew that freezes when the air temperature drops below freezing is:
   a. frost
   b. black frost
   c. hoarfrost
   d. white frost
   e. frozen dew

ANSWER: e

3. The cooling of the ground to produce dew is mainly the result of :
   a. conduction
   b. radiational cooling
   c. cooling due to the release of latent heat
   d. advection

ANSWER: b

4. For frozen dew to form:
   a. the dew point must be above freezing
   b. the minimum temperature must fall to freezing or below
   c. dew must form and then freeze
   d. all of the above

ANSWER: d

5. Suppose it is a winter night and at about 11 pm the air cools to the dew-point temperature and a thick radiation fog develops. If the air continues to cool during the night, in 5 hours the dew point temperature will probably:
   a. decrease as the air becomes drier
   b. decrease as the air becomes moister
   c. increase as the air becomes drier
   d. increase as the air becomes moister

ANSWER: a

6.   Particles that serve as surfaces on which water vapor may condense are called:
        a. hydrophobic nuclei
        b. nacreous nuclei
        c. condensation nuclei
        d. scud

ANSWER: c

7.   Frost forms when:
        a. objects on the ground cool below the dew point temperature
        b. the dew point is 32° F or below
        c. water vapor changes into ice without first becoming a liquid
        d. all of the above

ANSWER: d

8.   On a humid day, the _____ causes salty potato chips left outside in an uncovered bowl to turn soggy.
        a. repelling of water by hydrophobic condensation nuclei
        b. repelling of water by hygroscopic condensation nuclei
        c. attraction of water by hydrophobic condensation nuclei
        d. attraction of water by hygroscopic condensation nuclei

ANSWER: d

9.   Frost typically forms on the inside of a windowpane (rather than the outside) because
        a. the inside of the pane is colder than the outside
        b. there is more water vapor touching the inside of the pane
        c. there is less water vapor touching the inside of the pane

ANSWER: b

10.   At midnight in a shallow layer of air near the ground, the air temperature is 44 °F and the dew point temperature is 36 °F. If the air cools at a rate of 2 °F per hour until the dew point is reached, and at a rate of 1 °F per hour thereafter, then at 6:00 am:
        a. the temperature will be 34 °F and no condensation will be present
        b. the temperature will be 32 °F with frost and frozen dew present
        c. the temperature will be 32 °F and only frost will be present
        d. the temperature will be 32 °F and only frozen dew will be present
        e. the temperature will be 34 °F and only dew will be present

ANSWER: e

11.   Condensation nuclei may be:
        a. particles of dust
        b. nitric acid particles
        c. smoke from forest fires
        d. salt from the ocean
        e. all of the above

ANSWER: e

12.   Which of the following statements is(are) correct?
        a. the largest concentration of condensation nuclei are usually observed near the
           earth's surface
        b. wet haze restricts visibility more than dry haze
        c. fog is actually a cloud resting on the ground
        d. with the same water vapor content, fog that forms in dirty air is usually thicker
           than fog that forms in cleaner air
        e. all of the above are correct

ANSWER: e

13.   Condensation nuclei are important in the atmosphere because:
        a. they provide most of the minerals found in water
        b. they make it easier for condensation to occur in the atmosphere
        c. they are the sole producers of smog
        d. they filter out sunlight
        e. they provide the only means we have of tracking wind motions

ANSWER: b

14.   Under what circumstances could the relative humidity exceed 100 percent without producing fog?
        a. the dew point is higher than the air temperature
        b. the air is perfectly dry
        c. there are no condensation nuclei present
        d. there are too many electrically-charged ions in the air

ANSWER: c

15.   Wet haze forms when the relative humidity is:
        a. equal to 100%
        b. above 100%
        c. less than 100%
        d. equal to the dew point temperature

ANSWER: c

16.   Wet haze can form when the relative humidity is as low as ___ percent.
        a. 25
        b. 10
        c. 50
        d. 75
        e. 100

ANSWER: d

17.   When you see a layer of wet haze, you know that:
        a. the relative humidity of the air layer is 75 percent or more
        b. hygroscopic nuclei are present
        c. condensation is occurring on some nuclei
        d. all of the above

ANSWER: d

18.   On a cold winter morning the air near the surface is full of smoke particles. If fog should form in this air, it will probably be ___ fog that forms in cleaner air.
        a. thicker than
        b. thinner than
        c. practically the same as
        d. a different color than

ANSWER: a

19.   High fog is also called
        a. stratus
        b. cumulus
        c. cirrus
        d. frost

ANSWER: a

20.   When radiation fog "burns off", the fog tends to dissipate
        a. from the bottom up
        b. from the top down
        c. starting at the middle, and working both upward and downward

ANSWER: a

21. Radiation fog forms best on a:
    a. clear winter night with a slight breeze
    b. cloudy winter night with a strong breeze
    c. clear summer night with a strong breeze
    d. cloudy summer night with a slight breeze
    e. cloudy winter night with a slight breeze

ANSWER: a

22. When fog "burns off" it:
    a. absorbs sunlight and warms up
    b. evaporates
    c. thickens from the ground up
    d. settles to the ground in the form of rain

ANSWER: b

23. On a clear night, the minimum temperature drops to 34 °F. The following night, fog forms early in the evening. It is a good bet the minimum temperature will not be as low because of the:
    a. enhancement of the greenhouse effect by the fog cloud
    b. trapping of smoke and dust by the water vapor
    c. calm winds that accompany fog
    d. all of the above

ANSWER: a

24. Which statement(s) below is(are) correct?
    a. valleys are more susceptible to radiation fog than hill tops
    b. without the summer fog along the coast of California, redwoods would not
       grow well there
    c. fog can be composed of ice crystals
    d. all of the above are correct

ANSWER: d

25. Winter fog in the Central Valley region of California is mainly due to:
    a. adiabatic cooling
    b. radiational cooling
    c. advection of cold air
    d. uplift
    e. evaporation of water from rivers

ANSWER: b

26.   On a cold, winter morning the most likely place for radiation fog to form is:
a. at the top of a hill or mountain
b. in a valley
c. along the side of a hill
d. over a body of water

ANSWER: b

27.   "Arctic sea smoke" and "steam devils" are forms of this type of fog:
a. radiational fog
b. advection fog
c. evaporation (mixing) fog
d. upslope fog

ANSWER: c

28.   The fog that forms along the Pacific coastline of North America is mainly this type:
a. radiation fog
b. upslope fog
c. frontal fog
d. advection fog
e. steam fog

ANSWER: d

29.   Fog that forms off the coast of Newfoundland is mainly a form of:
a. advection fog
b. frontal fog
c. steam fog
d. radiation fog
e. upslope fog

ANSWER: a

30.   Along an irregular coastline, advection fog is more likely to form at the headlands than at the beaches because of the ___ surface winds and ___ air.
a. converging, sinking
b. diverging, sinking
c. converging, rising
d. diverging, rising

ANSWER: c

31.   If fog is forming at Denver, Colorado, and the wind is blowing from the east, then the fog is most likely:
        a. advection fog
        b. frontal fog
        c. upslope fog
        d. radiation fog

ANSWER: c

32.   Exhaled breath from your mouth can condense when
        a. it is very cold
        b. it is very warm and humid
        c. the addition of water vapor from your breath causes the air's relative humidity to exceed 100%
        d. all of the above

ANSWER: d

33.   When you see your breath on a cold morning the air temperature:
        a. must be above freezing
        b. must be below freezing
        c. can be above or below freezing
        d. must be equal to the relative humidity

ANSWER: c

34.   The use of helicopters to mix the air and disperse radiation fog at airports is effective provided that
        a. the fog layer is not deep
        b. the liquid water content of the fog is low
        c. the liquid water content of the fog is high
        d. both a and b
        e. none of the above

ANSWER: c

35.   On a cold, calm autumn morning the formation of fog above a relatively warm lake would most likely be:
        a. radiation fog
        b. steam fog
        c. frontal fog
        d. advection fog
        e. upslope fog

ANSWER: b

36.   Which fog does <u>not necessarily</u> form in air that is cooling?
       a. advection fog
       b. radiation fog
       c. evaporation (mixing) fog
       d. upslope fog

ANSWER: c

37.   Frontal fog most commonly forms as ___ raindrops fall into a layer of ___ air.
       a. cold, warmer
       b. cold, windy
       c. warm, colder
       d. warm, windy

ANSWER: c

38.   Fog that most often forms as warm rain falls into a cold layer of surface air is called:
       a. radiation fog
       b. evaporation (mixing) fog
       c. advection fog
       d. upslope fog

ANSWER: b

39.   Clouds are classified by their
       a. appearance
       b. altitude
       c. method of formation
       d. all of the above
       e. only b and c

ANSWER: a

40.   A reasonably successful method of dispersing cold fog is to:
       a. seed the fog with dry ice
       b. warm the air so that the fog evaporates
       c. mix the surface air with warmer air above
       d. use the turboclair technique

ANSWER: a

41.   Fog is a major hazard to aviation.
       a. true
       b. false

ANSWER: a

42. Which association below is not correct?
    a. cirrocumulus - high cloud
    b. cumulus - cloud of vertical extent
    c. altostratus - high cloud
    d. stratus - low cloud

ANSWER: c

43. At which city might you be able to observe cirrus clouds at an altitude of 3000 m (10,000 feet) above the surface?
    a. Barrow, Alaska
    b. Honolulu, Hawaii
    c. Miami, Florida
    d. Chicago, Illinois

ANSWER: a

44. Even at high elevations where cirrus clouds are found, liquid water still exists in the clouds.
    a. true
    b. false

ANSWER: a

45. Which association below is not correct?
    a. cumulus congestus - anvil top
    b. cumulus - fair weather cumulus
    c. altocumulus castellanus - resemble "little castles"
    d. stratus fractus - scud
    e. cumulonimbus - thunderstorm clouds

ANSWER: a

46. Which cloud is least likely to produce precipitation that reaches the ground?
    a. stratus
    b. nimbostratus
    c. cumulonimbus
    d. cirrocumulus

ANSWER: d

47. In middle latitudes, which cloud will have the highest base?
    a. cirrostratus
    b. cumulonimbus
    c. altostratus
    d. cumulus

ANSWER: a

48.   A halo around the moon means that:
        a. cirrostratus clouds are present
        b. the clouds overhead are low clouds
        c. rain is falling from the clouds overhead
        d. the clouds are composed of water droplets

ANSWER: a

49.   Which cloud type is composed of ice crystals and can cause a halo to form around the sun or moon?
        a. altostratus
        b. stratus
        c. nimbostratus
        d. cirrostratus
        e. angelitus

ANSWER: d

50.   Which of the following associations is <u>not</u> correct?
        a. altostratus - middle cloud
        b. cirrus - high cloud
        c. stratocumulus - cloud of vertical development
        d. cirrocumulus - high cloud
        e. cumulonimbus - cloud of vertical development

ANSWER: c

51.   Light or moderate-but-steady precipitation is most often associated with ___ clouds.
        a. nimbostratus
        b. cirrostratus
        c. cirrocumulus
        d. cumulonimbus

ANSWER: a

52.   In middle latitudes, which cloud will have the lowest base?
        a. cirrostratus
        b. stratocumulus
        c. altocumulus
        d. cirrus

ANSWER: b

53. Which of the following cloud types would be found at the highest elevation above the earth's surface?
    a. cumulonimbus
    b. cirrocumulus
    c. noctilucent
    d. cumulus congestus

ANSWER: c

54. Which of the following pairs of cloud types could be very similar in appearance?
    a. cumulus and cirrus
    b. cirrostratus and stratus
    c. altocumulus and cirrocumulus
    d. cirrocumulus and cumulonimbus

ANSWER: c

55. Which clouds often appear in parallel waves or bands?
    a. steam fog
    b. altocumulus
    c. cumulus congestus
    d. cirrostratus

ANSWER: b

56. A "mackerel sky" describes what type of cloud?
    a. cirrocumulus
    b. stratocumulus
    c. cumulonimbus
    d. nimbostratus
    e. cumulus

ANSWER: a

57. When viewed from the surface, the smallest individual cloud elements (puffs) are observed with which cloud?
    a. stratocumulus
    b. cumulus
    c. cirrocumulus
    d. altocumulus
    e. cumulonimbus

ANSWER: c

58.   Cirrus clouds are composed primarily of:
       a. water droplets
       b. water vapor
       c. ice particles
       d. salt aerosols

ANSWER: c

59.   Detached clouds of delicate and fibrous appearance, without shading, usually white in color and sometimes of a silky appearance are:
       a. stratus
       b. cirrocumulus
       c. altostratus
       d. cirrus

ANSWER: d

60.   Suppose the sky is completely covered with a thin, white layered-type cloud. You look at the ground and see that objects cast a distinct shadow. From this you conclude that the cloud type must be:
       a. stratus
       b. nimbostratus
       c. cirrostratus
       d. stratocumulus

ANSWER: c

61.   At middle latitudes, the base of an altostratus or altocumulus cloud would generally be found between:
       a. 200 and 6500 feet
       b. 6500 and 23,000 feet
       c. 23,000 and 43,000 feet
       d. above 43,000 feet

ANSWER: b

62.   A middle cloud that sometimes forms in parallel waves or bands is:
       a. cirrocumulus
       b. cumulonimbus
       c. altocumulus
       d. stratocumulus
       e. altostratus

ANSWER: c

63. A dim, "watery" sun visible through a gray sheet-like cloud layer is often a good indication of
    clouds.
    a. stratocumulus
    b. cirrostratus
    c. cumulonimbus
    d. altostratus
    e. nimbostratus

ANSWER: d

64. The name given to ragged-looking clouds that rapidly drift with the wind beneath a rain-
    producing cloud is:
    a. pileus
    b lenticular clouds
    c. castellanus
    d. scud

ANSWER: d

65. An anvil-shaped top is most often associated with:
    a. cumulonimbus
    b. cumulus congestus
    c. altocumulus
    d. cumulus humilis

ANSWER: a

66. If you hold your hand at arm's length and cloud elements appear to be about the size of your fist,
    the cloud type is probably:
    a. cumulus humilis
    b. altocumulus
    c. cirrocumulus
    d. stratocumulus

ANSWER: d

67. The name given to a towering cloud that has not fully developed into a thunderstorm is:
    a. cumulus humilis
    b. cumulus congestus
    c. cumulonimbus
    d. altocumulus

ANSWER: b

68.   A low, lumpy cloud layer that appears in rows, patches or rounded masses would be classified:
       a. nimbostratus
       b. stratus
       c. altocumulus
       d. stratocumulus

ANSWER: d

69.   Hail is usually associated with what cloud?
       a. stratus
       b. cumulus
       c. stratocumulus
       d. altocumulus
       e. cumulonimbus

ANSWER: e

70.   The cloud with the greatest vertical growth is:
       a. cumulus congestus
       b. cumulus humilis
       c. cumulonimbus
       d. cirrocumulus

ANSWER: c

71.   As Apollo 12 ascended into the atmosphere, the height of the surrounding clouds was noted to be 42,000 feet. A lightning stroke was seen within these clouds, indicating that they must have been:
       a. cumulus congestus
       b. cumulonimbus
       c. cirrus
       d. cirrocumulus
       e. lenticular

ANSWER: b

72.   Which of the following clouds would form in the stratosphere?
       a. cirrostratus
       b. nacreous
       c. lenticular
       d. mammatus

ANSWER: b

73. Which cloud forms in descending air?
    a. cumulus fractus
    b. cumulonimbus
    c. mammatus
    d. pileus

ANSWER: c

74. Which below is <u>not</u> a way in which a contrail may form?
    a. from water vapor in the engine exhaust mixing with air
    b. by air cooling as it passes over the aircraft's wings
    c. due to heating of the air by the engine exhaust
    d. all of the above

ANSWER: c

75. The small smooth cloud that may form just above the top of a towering cumulus cloud is called (a):
    a. mammatus cloud
    b. pileus cloud
    c. contrail
    d. banner cloud
    e. scud

ANSWER: b

76. The cloud-like streamer often seen forming behind an aircraft flying at high altitude is called (a):
    a. contrail
    b. pileus
    c. mammatus
    d. banner
    e. scud

ANSWER: a

77. Clouds that have a characteristic lens-shaped appearance are referred to as (a):
    a. lenticular
    b. mammatus
    c. contrail
    d. banner clouds

ANSWER: a

78.   Clouds that appear as bag-like sacks hanging from beneath a cloud are:
        a. pileus
        b. lenticular
        c. mammatus
        d. castellanus
        e. scud

ANSWER: c

79.   An altocumulus in the form of parallel waves would be described as an altocumulus ___.
        a. incus
        b. calvus
        c. translucidus
        d. capillatus
        e. undulatus

ANSWER: e

80.   Another name for a "mother of pearl" cloud:
        a. noctilucent cloud
        b. pileus cloud
        c. banner cloud
        d. nacreous cloud

ANSWER: d

81.   "Luminous night clouds" are also called:
        a. lenticular clouds
        b. noctilucent clouds
        c. pileus clouds
        d. nacreous clouds
        e. contrails

ANSWER: b

82.   Which term below describes a situation in which clouds cover between one-tenth and five-tenths
        of the sky?
        a. scattered
        b. broken
        c. overcast
        d. obscured

ANSWER: a

83.   When clouds are viewed near the horizon, the individual cloud elements usually:
       a. appear closer together than is actually the case
       b. appear farther apart than is actually the case
       c. appear lighter in color than is actually the case
       d. appear to have more vertical development than is actually the case

ANSWER: a

84.   Which of the following can be used to determine the height of cloud bases?
       a. ceiling balloons
       b. rotating-beam ceilometer
       c. fixed-beam ceilometer
       d. all of the above

ANSWER: d

85.   If a pilot balloon rises at a rate of 100 m per minute, and if it disappears into a deck of stratus clouds 1500 m (5000 ft) thick in 5 minutes, what is the ceiling of the cloud layer?
       a. 100 m (300 ft)
       b. 300 m (1000 ft)
       c. 500 m (1600 ft)
       d. 1500 m (5000 ft)
       e. 2000 m (6400 ft)

ANSWER: c

86.   Which of the following would provide the most accurate method of determining cloud base altitude?
       a. weather radar
       b. ceilometer
       c. geosynchronous satellite
       d. visual observation

ANSWER: b

87.   Infrared and visible satellite photographs might provide:
       a. a way of determining cloud thickness and altitude
       b. a way of distinguishing between wet and dry clouds
       c. a way of identifying clouds suitable for cloud seeding
       d. a way of distinguishing between "new" and "old" clouds

ANSWER: a

88. On an infrared satellite picture, low warm clouds appear ___ and high cold clouds appear ___.
    a. white, gray
    b. white, white
    c. gray, white
    d. gray, gray

ANSWER: c

89. Infrared satellite pictures are computer enhanced to:
    a. increase the contrast between specific features in the picture
    b. show where thick clouds with cold tops are located
    c. show where clouds with tops near the freezing level are located
    d. all of the above

ANSWER: d

90. Satellite photographs taken of clouds at night use:
    a. reflected visible light
    b. reflected infrared light
    c. emitted infrared light
    d. microwave radiation

ANSWER: c

91. If a cloud appears white on a visible satellite photograph and gray on an infrared picture, then the cloud could be:
    a. stratus
    b. cumulonimbus
    c. altostratus
    d. cirrostratus

ANSWER: a

92. If a cloud appears white in a visible satellite photograph and white in an infrared picture, then the cloud could be:
    a. cirrocumulus
    b. stratus
    c. cumulus humilis
    d. stratocumulus

ANSWER: a

93.   Which of the following is <u>not</u> a name given to a satellite?
        a. SCUD
        b. LandSat
        c. TIROS
        d. Nimbus
        e. GOES

ANSWER: a

94.   Geostationary satellites:
        a. orbit the earth once each day
        b. remain above a fixed spot above the equator
        c. are placed in higher orbits than most polar orbiting satellites
        d. all of the above

ANSWER: d

95.   Satellites can:
        a. monitor the amount of snow cover
        b. provide information about the earth-atmosphere energy balance
        c. provide information about surface water temperatures
        d. monitor the movement of icebergs
        e. all of the above

ANSWER: e

96.   Polar orbiting satellites:
        a. remain above a fixed point over the North or South Pole
        b. pass directly over the same place on each orbit
        c. on each successive orbit view an area to the west of the previous orbit
        d. follow a line of constant latitude around the earth

ANSWER: c

97.   A rotating-beam ceilometer is useful for
        a. estimating the altitude of the tops of vertical development clouds
        b. detecting the presence of haze
        c. determining visibility for pilots

ANSWER: c

98.   Lenticular clouds typically form _____ a mountain range.
        a. downwind of
        b. upwind of
        c. far away from
        d. about 10 km above

ANSWER: a

Essay Exam Questions

1. It is a cold winter night and a fog cloud forms. If it continues to cool during the night would you expect to find that the dew point has changed overnight? If so, would you expect to find that dew point has increased or decreased?

2. List the main types of fog, then briefly explain how each one forms. Where might you expect each of these different types of fog to form?

3. List one or more key identifying features for each of the ten basic cloud types. Which cloud types might have fairly similar appearances and thus be difficult to identify?

4. With which cloud type would each of the following phenomena be associated?
   lightning   hail   tornadoes
   brief heavy rain   steady rain   milky sun
   halo   mackerel sky

5. Suppose the sky is covered with stratus clouds. How might you determine whether middle or high clouds are also present?

6. A vertically-viewing detector is located 8000 feet from a rotating beam ceilometer. Determine the height of the cloud base assuming a signal is detected when the ceilometer beam elevation angle is 30°.

7. List the major height categories of clouds. What differences might you expect to find in the clouds that form at these different levels?

8. Fog is often described as a cloud that forms at ground level. In what ways are the formation of fog and clouds similar and different?

9. Explain why clouds form in rising air. Would it be possible for rising air to remain cloud-free?

10. Describe, with the aid of a figure, the orbital paths followed by geostationary and polar-orbiting satellites.

11. What kinds of information about clouds could you hope to determine using infrared and visible satellite photographs?

12. List and discuss some of the effects that you think weather satellites may have had on the field of weather prediction.

13. Under what conditions can a cloud form when air is sinking?

Chapter 7
Stability and Cloud Development

Summary

This chapter examines atmospheric stability and factors that affect the development of clouds. The concept of stable and unstable equilibria is introduced using the familiar, instructive analogy of a rock placed on a hilltop or in a valley. The student is shown that stability in the atmosphere depends on the change of temperature in a moving parcel relative to its surroundings. In a stable atmosphere, a rising parcel will become colder and denser than its surroundings and will resist further upward motion. Clouds which form in a stable atmosphere tend to develop horizontally and have a layered structure. In an unstable region of the atmosphere, a rising air parcel will become warmer and less dense than its surroundings and will continue to move upward on its own, often forming cumuliform clouds. Some of the conditions that affect or may change atmosphere stability, such as warming or cooling at the ground, the influx of warm or cold air aloft, and the upward or downward motion of an air layer, are discussed. In a focus section, a mathematical model for estimating the base of convective clouds is presented.

The rising air motions that are needed to form clouds can be produced in a variety of ways including convection, topographic lifting, convergence, and uplifting at frontal boundaries. Several examples of particular cloud formations and the upper level conditions that produce them are examined.

A lengthy focus section on adiabatic charts presents the notion of graphical solutions to thermodynamic problems, a familiar concept to many professional meteorologists.

Key Terms

stable equilibrium
unstable equilibrium
parcel of air
adiabatic process
dry adiabatic rate
moist adiabatic rate
pseudoadiabatic
sounding
lapse rate
environmental lapse rate
absolutely stable
  atmosphere
subsidence inversion
neutral stability
absolutely unstable
  atmosphere

autoconvective lapse
  rate
conditionally unstable
  atmosphere
conditional instability
convective instability
thermal
convection current
condensation level
entrainment
orographic uplift
orographic clouds
rain shadow
lifting condensation
  level (LCL)
lenticular clouds

mountain wave clouds
lee wave clouds
standing wave clouds
rotor clouds
adiabatic chart
dry adiabats
potential temperature
moist adiabats
mixing ratio
cloud streets
billow clouds
altocumulus castellanus
cirrocumulus castellanus

Teaching Suggestions

1.   Free convection can be demonstrated by filling a dark-colored balloon with hydrogen or helium gas. Tape a small piece of paper to the balloon to add weight. Trim the paper until the balloon is just heavy enough that it will sink very slowly when released. Heat the balloon using an infrared heat lamp or a high-wattage flood light. As the balloon warms, it will expand slightly and begin to rise. The balloon will float out of range of the heat lamp, cool and sink back toward the floor. The demonstration can be repeated several times.

2.   An interesting class discussion can sometimes be generated by writing two statements, "rising air cools" and "warm air rises", on the board and asking the students whether the statements are contradictory.

3.   This chapter might be an appropriate place to introduce Archimedes' Law. Students should be able to understand that the vertical pressure gradient force and the downward pull of gravity are generally very nearly balanced in the atmosphere. A slight change in an air parcel's density (with respect to its surroundings) will upset the balance and cause the parcel to rise or sink. For a surprising demonstration of density differences, place a can of diet and sugar-sweetened soft drink in an aquarium filled with water. The can of diet drink will generally float, while the sweetened beverage will sink. A large amount of corn syrup is added to soft drinks and the resulting mixture has a density greater than water.

4.   Satellite cloud photographs often illustrate stable and unstable atmospheric conditions. A spotty cumulus cloud pattern is often visible when cold air moves out over warm ocean water, for example.

5.   Ask students to explain why, when they've depressed the valve on an inflated bicycle tire, their finger gets cold.

Student Projects

1.   Strong radiation inversions frequently develop overnight in many locations. Have students plot an early morning sounding and identify the top of the inversion layer. In larger cities, an associated hazy layer may be visible over the city in the early morning. Plot the afternoon sounding for the same day and determine whether the inversion layer has disappeared. Are any other inversion layers visible? Have the students investigate what meteorological conditions favor strong inversions and pollution episodes for their location. (Tabulated sounding data will generally use pressure as the vertical coordinate; in this case it is probably sufficient to assume a 1 mb decrease per 10 meters)

Blue Skies   2.   Use the Atmospheric Basics/Layers of the Atmosphere section of the Blue Skies cdrom to explore the vertical profile of temperature at a location near you. What is the environmental lapse rate of the first 500 meters of the atmosphere? What is the stability of this layer?

Blue Skies   3.   Using the Moisture and Stability/Adiabatic activity on the BlueSkies cdrom, make a graph of the altitude of cloud formation (plot this on the vertical scale) vs. dew point depression (plotted on the horizontal scale). Dew point depression is the difference between the temperature and the dew point temperature. Explain the shape of the graph.

Answers to Questions for Thought

1.   When a layer of air is absolutely stable the environmental lapse rate is less than the moist adiabatic rate. However, if the top of the layer is unsaturated and the bottom of the layer is saturated, lifting the entire layer will change its stability to one of instability.

2.   Generally, at night the spread between air temperature and dew point is less than during the day. Consequently, air need not rise as high at night to become saturated. Hence, if convection does occur (such as over warm water) the cloud bases are often lower than they would be during the day.

3.   Along a deep valley. The tall mountainsides along either side of the valley should shelter the airport from dangerous cross breezes. Nighttime landings may be hazardous, though, when drainage winds are present.

4.   The actual vapor pressure of the air decreases as we move west of a line that runs through central Kansas. This means that, if the air temperature remains the same, there will be a greater spread between air temperature and dew point and, hence, a higher cloud base.

5.   In the afternoon when the surface air temperature is highest and the air is most unstable.

6.   Relative humidity depends on temperature, in addition to the amount of water vapor in the air. As air descends, its temperature increases, thus decreasing the relative humidity.

Answers to Problems and Exercises

1.   The cumulus cloud bases in (e) would be observed at the highest level, about 3250 m above the surface (rising air cools at 10 °C/km until saturation, the dew point temperature decreases at a rate of 2 °C/km with increasing altitude).

2.   Air temperature at surface = 28 °C. In rising air, the air parcel will cool 10 °C to 18 °C. The dew point temperature will decrease 2 °C to 18 °C.

3.   The difference between air temperature and dew point at the surface is 9 °F. If we assume a dew point lapse rate of 1 °F per 1000 feet, then the surface dew point must be 73 °F + 2 °F or 75 °F. If the separation between air temperature and dew point is 9 °F, then the surface air temperature must be about 84 °F. Using saturation water vapor pressure data in Chap. 5 we find e = 29.6 mb, e_s = 40 mb, and the RH = 75%.

4.   -50 °C + 10 km x 10 °C/km = 50 °C (122 °F)

5.   (a) 8 °C/1000 m
        (b) conditional instability
        (c) RH = 100% x e/e_s = 100% x (6.9 mb)/(12.3 mb) = 56%
        (d) 1000 m
        (e) 0 °C
        (f) -12 °C
        (g) Air inside cloud is 2 °C warmer than the surrounding air; this suggest unstable air.
        (h) Air should rise because it is warmer and less dense than the air surrounding it.
        (i) cumulus congestus (dew point is -12 °C on the mountain top and increases 2 °C for every kilometer of descent.
        (j) temperature = 18 °C, dew point = -6 °C
        (k) RH = 100% x e/e_s = 100% x (4.0 mb)/(21 mb) = 19%
        (l) Latent heat released during condensation on western (windward) side.
        (m) Water vapor removed from air during condensation and precipitation on western side.

Multiple Choice Exam Questions

1.   If the environmental lapse rate is 5° C per 1000 m and the temperature at the earth's surface is 25° C, then the air temperature at 2000 m above the ground is:
        a. 25° C
        b. 30° C
        c. 20° C
        d. 15° C

ANSWER: d

2. If a parcel of unsaturated air with a temperature of 30° C rises from the surface to an altitude of 1000 m, the unsaturated parcel temperature at this altitude would be about:
    a. 10° C warmer than at the surface
    b. 10° C colder than at the surface
    c. 6° C colder than at the surface
    d. impossible to tell from the data given

ANSWER: b

3. If an air parcel is given a small push upward and it falls back to its original position, the atmosphere is said to be:
    a. stable
    b. unstable
    c. isothermal
    d. neutral
    e. adiabatic

ANSWER: a

4. The rate at which the actual air temperature changes with increasing height above the surface is referred to as the:
    a. environmental lapse rate
    b. dry adiabatic rate
    c. moist adiabatic rate
    d. thermocline

ANSWER: a

5. A rising parcel of air that does not exchange heat with its surroundings is an example of
    a. isothermal ascent
    b. an adiabatic process
    c. forced lifting
    d. advection

ANSWER: b

6. The rate at which the temperature changes inside a rising (or descending) parcel of saturated air is called the:
    a. environmental lapse rate
    b. dry adiabatic lapse rate
    c. moist adiabatic lapse rate
    d. latent heat release rate

ANSWER: c

7.   At the earth's surface, a rising saturated air parcel would cool most rapidly when its temperature is:
a. 10 °F
b. 32 °F
c. 50 °F
d. 68 °F
e. 80 °F

ANSWER: a

8.   The dry adiabatic rate is the rate at which:
a. an air parcel rises
b. temperature changes in a rising or descending parcel of unsaturated air
c. volume changes when a parcel expands or is compressed
d. latent heat energy is released in a rising air parcel

ANSWER: b

9.   A completely dry air parcel which first rises and cools, and subsequently sinks and warms, is undergoing
a. an irreversible pseudoadiabatic process
b. a reversible adiabatic process
c. an irreversible adiabatic process

ANSWER: b

10.   A sounding can be measured by
a. a thermometer
b. a psychrometer
c. a radiosonde
d. a barograph

ANSWER: c

11.   The difference between the "moist" and "dry" adiabatic rates is due to:
a. the fact that saturated air is always unstable
b. the fact that an unsaturated air parcel expands more rapidly than a saturated air parcel
c. the fact that moist air weighs less than dry air
d. the fact that latent heat is released by a rising parcel of saturated air

ANSWER: d

12.   The dry adiabatic lapse rate is _____ greater than the moist adiabatic lapse rate.
        a. never
        b. sometimes
        c. always

ANSWER: c

13.   A knowledge of air stability is important because:
        a. it determines the direction of movement of storms
        b. it determines the vertical motion of air
        c. it determines the movement of high pressure areas
        d. it determines seasonal weather patterns

ANSWER: b

14.   The most latent heat would be released in a __ parcel of __ saturated air.
        a. rising, warm
        b. rising, cold
        c. sinking, warm
        d. sinking, cold

ANSWER: a

15.   Most thunderstorms do not extend very far into the stratosphere because the air in the stratosphere
        is:
        a. unstable
        b. stable
        c. too cold
        d. too thin
        e. too dry

ANSWER: b

16.   Which set of conditions, working together, will make the atmosphere the most stable?
        a. cool the surface and warm the air aloft
        b. cool the surface and cool the air aloft
        c. warm the surface and cool the air aloft
        d. warm the surface and warm the air aloft

ANSWER: a

17.   Which cloud type would most likely form in absolutely stable air?
        a. cumulus congestus
        b. cumulonimbus
        c. stratus
        d. altocumulus

ANSWER: c

18.   Subsidence inversions are best developed with ___ pressure areas because of the ___ air motions associated with them.
        a. high, rising
        b. high, sinking
        c. low, rising
        d. low, sinking

ANSWER: b

19.   An inversion represents an extremely stable atmosphere because air that rises into the inversion will eventually become ___ and ___ dense than the surrounding air.
        a. warmer, less
        b. warmer, more
        c. colder, less
        d. colder, more

ANSWER: d

20.   If the environmental lapse rate is less than the moist adiabatic rate, the atmosphere is:
        a. conditionally unstable
        b. absolutely stable
        c. absolutely unstable
        d. neutrally stable

ANSWER: b

21.   Which of the following conditions would be described as the most stable?
        a. environmental lapse rate is 13 °C per kilometer
        b. environmental lapse rate is 3 °C per kilometer
        c. isothermal conditions
        d. an inversion

ANSWER: d

22.   Which of the following environmental lapse rates would represent the most unstable atmosphere in a layer of unsaturated air?
        a. 3° C per 1000 m
        b. 6° C per 1000 m
        c. 9° C per 1000 m
        d. 11° C per 1000 m

ANSWER: d

23. In a conditionally unstable atmosphere, the environmental lapse rate will be ___ than the moist adiabatic rate and ___ than the dry adiabatic rate.
    a. greater, less
    b. greater, greater
    c. less, greater
    d. less, less

ANSWER: a

24. If an air parcel is given a small push upward and it continues to move upward on its own accord, the atmosphere is said to be:
    a. stable
    b. unstable
    c. buoyant
    d. dynamic

ANSWER: b

25. A conditionally unstable atmosphere is ___ with respect to unsaturated air and ___ with respect to saturated air.
    a. unstable, stable
    b. unstable, unstable
    c. stable, unstable
    d. stable, stable

ANSWER: c

26. When the environmental lapse rate decreases more rapidly with height than the dry adiabatic rate, the atmosphere is:
    a. absolutely stable
    b. absolutely unstable
    c. convectively unstable
    d. conditionally unstable

ANSWER: b

27. What two sets of conditions, working together, will make the atmosphere the most unstable?
    a. cool the surface and warm the air aloft
    b. cool the surface and cool the air aloft
    c. warm the surface and cool the air aloft
    d. warm the surface and warm the air aloft

ANSWER: c

138

28. Which condition below could make a layer of air more unstable?
a. mix the air in the layer
b. lift the entire air layer
c. cool the upper part of the layer
d. all of the above

ANSWER: d

29. Convective instability associated with severe thunderstorms and tornadoes can be brought on by:
a. lifting a stable air layer whose surface is humid and whose top is dry
b. forcing a layer of moist air to descend a mountain range
c. heating the upper portion of an unsaturated layer of air
d. all of the above

ANSWER: a

30. If unsaturated stable air is lifted to a level where it becomes saturated and unstable, this type of instability is called:
a. conditional instability
b. convective instability
c. baroclinic instability
d. forced instability

ANSWER: a

31. Which cloud type below would most likely form in an unstable atmosphere?
a. cumulonimbus
b. stratus
c. cirrostratus
d. nimbostratus
e. cumulus humilis

ANSWER: a

32. Just above cumulus humilis clouds you would expect to find:
a. a stable layer
b. an unstable layer
c. a conditionally unstable layer
d. unusually strong horizontal winds

ANSWER: a

33.   Which of the following set of surface conditions would produce a convective cumulus cloud with the highest base?
      a. air temperature 40 °C, dew point 30 °C
      b. air temperature 25 °C, dew point 10 °C
      c. air temperature 35 °C, dew point 10 °C
      d. air temperature 45 °C, dew point 25 °C
      e. air temperature 25 °C, dew point 5 °C

ANSWER: c

34.   _____ are found beneath fair-weather cumulus (cumulus humilis) clouds, whereas _____ are found between the clouds.
      a. downdrafts; updrafts
      b. updrafts; downdrafts
      c. downdrafts; only horizontal motions
      d. updrafts; only horizontal motions

ANSWER: b

35.   Which of the following sets of conditions would produce a cumulus cloud with the lowest base?
      a. air temperature 90° F, dew point temperature 50° F
      b. air temperature 90° F, dew point temperature 40° F
      c. air temperature 90° F, dew point temperature 60° F
      d. air temperature 90° F, dew point temperature 20° F

ANSWER: c

36.   In a rising thermal, the dew-point temperature:
      a. increases at a slower rate than the parcel temperature
      b. decreases at the same rate as the parcel temperature
      c. decreases at a slower rate than the parcel temperature
      d. increases at a faster rate than the parcel temperature

ANSWER: c

37.   Which of the following statements is (are) correct?
      a. convection can occur over the ocean
      b. air motions are usually downward around a cumulus cloud
      c. the temperature of the rising air at a given level inside a cumulus cloud is
         normally warmer that the air around the cloud
      d. all of the above

ANSWER: d

38. Which of the following is <u>not</u> a way of producing clouds?
    a. lifting air along a topographic barrier
    b. lifting air along a front
    c. warming the surface of the earth
    d. convergence of surface air
    e. air motions caused by subsidence

ANSWER: e

39. The vertical motion of air caused by sun heating the ground is called:
    a. convection
    b. orographic lifting
    c. subsidence
    d. convergence

ANSWER: a

40. If you were to take a trip during the summer from Ohio to Nevada, you would most likely observe that afternoon cumulus clouds ___ as you travel west.
    a. have higher cloud bases
    b. are larger
    c. form earlier in the day
    d. disappear

ANSWER: a

41. What would be the height of the base of a cumulus cloud when the surface air temperature is 45 °C and the dew point is 25 °C?
    a. 1125 m
    b. 2000 m
    c. 2500 m
    d. 4000 m
    e. 6500 m

ANSWER: c

42. Suppose the surface air temperature is 66 °F. If the base of a cumulus congestus cloud directly above you is 2000 feet, what would be the approximate temperature at an altitude of 5000 feet above you inside the cloud? (Hint: the dry adiabatic rate is 5.5 °F per 1000 feet, the moist adiabatic rate is 3.0 °F per 1000 feet)
    a. 55 °F
    b. 49 °F
    c. 46 °F
    d. 41 °F
    e. 32 °F

ANSWER: c

43.   What would the air temperature inside a conventional jet airliner be, if outside air at an altitude of 10 km, pressure of 250 mb and a temperature of -60 °C, is brought inside and compressed to a 1000 mb pressure? (Hint: you may assume that 1000 mb pressure is equivalent to 0 m altitude)
    a. -60 °C
    b. 0 °C
    c. 40 °C
    d. 60 °C

ANSWER: c

44.   Soaring birds in flight, like hawks and eagles, often indicate the presence of
    a. irreversible pseudoadiabatic processes
    b. absolute instability
    c. thermals
    d. subsidence inversions

ANSWER: c

45.   Standing wave clouds are usually a form of
    a. fog
    b. low cloud
    c. middle cloud
    d. high cloud

ANSWER: c

46.   An example of orographic clouds would be:
    a. clouds forming over a warm ocean current
    b. clouds forming on the windward slope of a mountain
    c. clouds forming behind a jet airplane
    d. clouds formed by surface heating

ANSWER: b

47.   An example of orographic clouds would be:
    a. cumulus clouds produced by surface heating
    b. rotor clouds
    c. lee-wave clouds
    d. clouds that form on the upwind slope of a mountain
    e. pileus cloud

ANSWER: d

48.   Which of the following cloud types would commonly be found downwind of a mountain:
       a. lenticular clouds
       b. anvil clouds
       c. mammatus clouds
       d. contrails

ANSWER: a

49.   Wave clouds that often form over or downwind of a mountain are called:
       a. lenticular clouds
       b. castellanus clouds
       c. mammatus clouds
       d. nacreous clouds
       e. contrails

ANSWER: a

50.   An adiabatic chart is a useful tool for determining
       a. a station model
       b. isobars
       c. the wind speed
       d. the lifting condensation level

ANSWER: d

51.   The name commonly used to describe the drier region observed on the downwind (leeward) side
       of a mountain range is:
       a. orographic
       b. inversion region
       c. rain shadow
       d. compression region

ANSWER: c

52.   One would most likely expect to see a rotor cloud:
       a. on top of a developing cumulus cloud
       b. in the middle of a cumulonimbus
       c. beneath a lenticular cloud
       d. in absolutely unstable air
       e. in the middle of a subsidence inversion

ANSWER: c

53. The temperature an air parcel would have if it were moved to a pressure of 1000 mb at the dry adiabatic rate is called the :
    a. descending temperature
    b. adiabatic temperature
    c. potential temperature
    d. dew point temperature
    e. base temperature

ANSWER: c

54. When altocumulus clouds become arranged in rows, the clouds are given the name:
    a. humilis
    b. rotors
    c. cloud streets
    d. castellanus
    e. pileus

ANSWER: c

55. Clouds such as altocumulus and cirrocumulus that develop vertical, tower-like extensions are called:
    a. rotors
    b. pileus
    c. billows
    d. lenticular
    e. castellanus

ANSWER: e

56. Which condition below is necessary for a layer of altostratus clouds to change into altocumulus?
    a. the top part of the cloud deck cools while the bottom part warms
    b. temperature increases with increasing altitude in the layer
    c. the cloud layer becomes more stable
    d. all of the above

ANSWER: a

57. Wave-like clouds that often appear as water breaking along a shore are called:
    a. cloud streets
    b. pileus clouds
    c. mammatus clouds
    d. billows
    e. rotors

ANSWER: d

58. Which process below will change a most layer of stable air into a deck of low stratocumulus clouds?
    a. mixing
    b. lifting
    c. cooling the surface
    d. heat the top of the layer
    e. subsidence

ANSWER: a

59. In a sky filled with scattered cumulus humilis clouds, rising motions are found _____ the clouds and sinking motions are found _____ the clouds.
    a. between; under
    b. under; between
    c. under; under
    d. between; between

ANSWER: b

Essay Exam Questions

1. An air parcel that is warmer than its surroundings will rise. What force accounts for this upward motion?

2. Explain how the stability of the atmosphere can affect the types of clouds that form.

3. Would you expect to find a subsidence inversion to be associated with high or low pressure? What effects might a subsidence inversion have on weather conditions at the ground?

4. Does radiational cooling at the ground at night act to increase or decrease atmospheric stability? How does daytime heating at the ground during the day affect atmospheric stability?

5. Based on atmospheric stability considerations, do you think it would be best to burn agricultural debris in the early morning or the afternoon?

6. Explain how lifting an air layer can steepen the environmental lapse rate and make the layer more unstable.

7. List and explain the four processes that cause rising air motions in the atmosphere.

8. Explain why the surface air on the downwind side of a mountain can be drier than the surface air on the upwind side. What is this effect called? Can you think of a location in the United States where this might actually occur?

9. Suppose you observe lenticular clouds that appear to be motionless. Is the air in the vicinity of the cloud motionless? Explain.

10. What types of clouds might you expect to see form when a cold mass of air moves over warmer water?

11. Cumulonimbus clouds indicate unstable atmospheric conditions. What stops the upward growth and causes the top of a cumulonimbus cloud to flatten out into an anvil?

12. Describe the changes in atmospheric stability throughout a 24 hour period beginning at sunrise on a cloudless summer day over land.

Chapter 8
Precipitation

Summary

This chapter examines the processes that produce precipitation and looks at the different types of precipitation that can fall from clouds.

The chapter begins with a more detailed look at the formation and growth of cloud droplets. Cloud droplets or ice crystals are themselves too small and light to be able to reach the ground as precipitation. Raindrops can form rapidly in warm clouds, however, when water droplets collide and coalesce. Formation of rain by this process works best in thick clouds which have strong updrafts. In cold clouds it is possible for ice crystals to grow when surrounded by supercooled water droplets. Attempts to enhance precipitation by cloud seeding are reviewed. A focus section discusses the process of artificially enhancing precipitation growth by cloud seeding.

Precipitation can reach the ground in a variety of forms depending on the type of cloud producing it and also on the atmospheric conditions between the cloud base and the ground. Finally, students learn how rain and snowfall amounts can be measured using simple instruments or estimated remotely using Doppler radar.

Key Terms

precipitation
saturation vapor
  pressure
equilibrium vapor
  pressure
curvature effect
supersaturation
condensation nuclei
hygroscopic nuclei
solute effect
terminal velocity
coalescence
cold clouds
supercooled (water
  droplets)
mixed clouds
glaciated (cloud)
homogeneous freezing
ice embryo
ice nuclei
deposition nuclei
freezing nuclei
contact freezing

contact freezing
contact nuclei
ice crystal (Bergeron)
  process
diffusion
accretion
riming
graupel
aggregation
snowflake
cloud seeding
rain
drizzle
virga
shower (rain)
cloudburst
acid rain
snow
fall streaks
dendrite
flurries (of snow)
snow squall
blizzard

sleet (snow pellets)
freezing rain
glaze
silver thaw
freezing drizzle
rime
rime ice
snow grains
snow pellets
soft hail
hailstones
hailstreak
standard rain gauge
trace (of
  precipitation)
tipping bucket
  rain gauge
weighing-type rain
  gauge
water equivalent
radar
Doppler radar

Teaching Suggestions

1.   Measure and compare the fall speeds of different size drops (fine mist, drops from an eye dropper, small and large water balloons).

2.   Have the class design a precipitation-enhancement experiment. What method for enhancing precipitation will be used? What materials will be needed? How will the experiment's success be assessed?

 Blue Skies 3.   Use the Atmospheric Basics/Layers of the Atmosphere section of the BlueSkies cdrom to explore a sounding from an area where precipitation is occurring. Which layer is most likely producing the precipitation?

Student Projects

1.   Rainfall measurement, the construction of simple and inexpensive rain gauges, and student projects have been discussed by J.T. Snow and S.B. Harley (ref: "Basic Meteorological Observations for Schools: Rainfall," *Bull. Amer. Meteorol. Soc., 69,* 498-507, 1988).

148

 2.   Use the Atmospheric Basics/Layers of the Atmosphere activity on the BlueSkies cdrom to find an upper-air sounding (graph of temperature vs. altitude) that shows the necessary conditions for sleet or freezing rain.

 3.   Use the Weather Forecasting/Forecasting section of the BlueSkies cdrom to locate regions of the country where precipitation is being reported. What is the relative humidity at these locations? Explain why the relative humidity may not be 100 percent even though the precipitation is occurring.

Answers to Question for Thought

1.   Large ice crystals fall with their flat surfaces parallel to the ground. This produces a large flat surface area-to-weight ratio and, hence, a large amount of air resistance which reduces the ice crystal's rate of fall.

2.   The turbulent vertical motions and towering extent of a warm cumulus cloud will accelerate the collision-coalescence process of producing rain. In a cold stratus cloud vertical motions are small, the liquid water content is low, and the collision-coalescence process is not as effective in initiating precipitation, especially when the air temperature is quite low.

3.   This is an example of the curvature effect. The relative humidity in a cloud is measured with respect to a flat water surface. When the air is saturated in a cloud (relative humidity equals 100 percent), it is unsaturated with respect to a curved droplet of pure water. The droplet of pure water is not in equilibrium so it evaporates.

4.   In this example, the most important process would be the ice crystal process because the collision-coalescence process requires that cloud droplets be of varying size so that drops will fall at different speeds.

5.   One reason appears to be that in clouds that form over land there are larger concentrations of nuclei than in clouds that form over water. Hence in clouds forming over land there are more, but smaller, cloud droplets. Because the clouds that form over water usually contain fewer nuclei, they contain larger droplets and a wider distribution of droplet sizes. This enhances the collision-coalescence process and makes these clouds more efficient at producing rain.

6.   The blizzard occurs about 4 km (13,000 ft) above the surface in the middle of a violent thunderstorm.

7.   This snowfall pattern could be the result of waves forming in the upper troposphere as air flows from west to east over the mountains. Cirriform clouds form in the rising part of the wave (wave crest) and seed ice crystals into a lower supercooled cloud layer, enhancing precipitation. The wave crests are probably located above Denver, and above the region 150 km east of Denver.

8.   A large drop has greater surface area than a small drop, thus the frictional drag of the air it's falling through may tear it apart into smaller drops.

9.   Lead has a deleterious effect on the human body, especially the nervous system and kidneys.

10.   "Holes" are occasionally produced in altocumulus clouds when cirriform clouds are above them. Ice crystals fall from the cirriform clouds and mix with the supercooled cloud droplets of the altocumulus, converting many water droplets into ice crystals. The ice crystals grow larger by the ice crystal process and eventually fall from the altocumulus, leaving the appearance of a hole in the cloud. Another possibility is that an aircraft has penetrated the altocumulus and either engine exhaust or vibrations (or both) have changed the supercooled cloud droplets into ice.

11.   If it is raining heavily and there is a strong temperature inversion, freezing rain can fall when surface temperatures are -12°C. The presence of a strong inversion ensures that cloud temperatures will be much higher than surface temperatures, thus the precipitation is rain rather than snow. Heavy rain (large drops) means large fall velocities, thus the raindrops become supercooled (but don't freeze) as they fall toward the ground.

12.   When snow becomes mixed with sleet, it often indicates that warm air aloft has moved into the region, causing the snowflakes to, at least partially, melt and then refreeze in the colder surface air. Frequently this is the first indication that warm air is moving into the region. Often, continued influx of warm air, first aloft, then at the surface, will raise the air temperature to the point that the snowflakes melt and become raindrops.

13.   Possible reasons include: the snow partially melted; blowing and drifting; the water content of the snow changed over time; measurement error; the snow partially sublimated.

Answers to Problems and Exercises

1.   13.2 cm (5.2 in.)

2.   900 times faster (refer to Table 8.1)

3.   (a) 23.8 minutes
        (b) 12,310 seconds = 205 minutes or 3.4 hours

4.   (a) 167 seconds or 2.8 minutes
        (b) slightly elongated and flattened at the bottom
        (c) cumuliform, cumulus congestus or cumulonimbus

5.   The rime forms in the very cold part of the cloud as the droplets freeze on the hailstone. The clear ice probably forms in the lower (wetter) half of the cloud. The hailstone in the diagram probably formed in the lower, wetter region. Caught in an updraft it rose into the cold (drier) region where it collected rime. It then descended collecting clear ice and, again, rose to higher levels where it collected a coating of rime. On its way down it collected a final coating of clear ice.

Multiple Choice Exam Questions

1. Which below best describes the solute effect?
    a. keeps water droplets from freezing at temperatures below 32 °F.
    b. removal of pollutants from the atmosphere by cloud droplets
    c. water droplets dissolve hygroscopic nuclei and condensation can occur at
       relative humidities less than 100 percent
    d. evaporation of cloud droplets and grow of ice crystals in a cold cloud

ANSWER: c

2. Which statement below best describes the curvature effect?
    a. Large cloud droplets fall faster than small droplets
    b. small droplets evaporate more quickly than large droplets
    c. small droplets collide and coalesce more easily than larger droplets
    d. explains the six-sided shape of ice crystals

ANSWER: b

3. Condensation onto hygroscopic nuclei is possible at relative humidities less than 100 percent due
    to the:
    a. curvature effect
    b. electrical charge on these nuclei
    c. solute effect
    d. crystalline structure of these nuclei

ANSWER: c

4. Which of the following is not an important factor in the production of rain by the collision-
    coalescence process?
    a. the updrafts in the cloud
    b. relative size of the droplets
    c. the number of ice crystals in the cloud
    d. cloud thickness
    e. the electric charge of the droplets

ANSWER: c

5. Which cloud type below will only produce precipitation by the collision-coalescence process?
    a. a thick, cold nimbostratus cloud
    b. a thick, warm cumulus cloud
    c. a thick, cold cumulus cloud
    d. a thick, supercooled cumulonimbus cloud with abundant nuclei
    e. a supercooled cumulus congestus cloud

ANSWER: b

6.   Large raindrops fall ____ than smaller raindrops, and have a ____ terminal velocity than small raindrops.
        a. faster, lesser
        b. faster, greater
        c. slower, lesser
        d. slower, greater

ANSWER: b

7.   Which cloud would most likely produce drizzle?
        a. stratus
        b. cumulus
        c. cumulus congestus
        d. cirrostratus
        e. cumulonimbus

ANSWER: a

8.   The merging of liquid cloud droplets by collision is called:
        a. coalescence
        b. riming
        c. accretion
        d. deposition

ANSWER: a

9.   If you observe large raindrops hitting the ground, you could probably say that the cloud overhead was ___ and had ___ updrafts.
        a. thick, weak
        b. thick, strong
        c. thin, weak
        d. thin, strong

ANSWER: b

10.   If rain falls on one side of a street and not on the other side, the rain most likely fell from a:
        a. nimbostratus cloud
        b. stratus cloud
        c. cumulonimbus cloud
        d. altostratus cloud
        e. altocumulus cloud

ANSWER: c

11.   During the ice crystal process of rain formation:
        a. only ice crystals need be present in a cloud
        b. ice crystals grow larger at the expense of the surrounding liquid cloud droplets
        c. the temperature in the cloud must be -40° C (-40° F) or below
        d. the cloud must be a cumuliform cloud

ANSWER: b

12.   The temperature at which you would expect a cloud to become completely glaciated is:
        a.   0 °C (32 °F)
        b. -5 °C (23 °F)
        c. -18 °C (0 °F)
        d. -40 °C (-40 °F)

ANSWER: d

13.   Homogeneous nucleation occurs when
        a. water vapor condenses onto hygroscopic nuclei
        b. water vapor condenses onto hydrophobic nuclei
        c. water vapor condenses without nuclei
        d. all the condensation nuclei are exactly the same

ANSWER: c

14.   When only ice crystals exist in a cloud, the cloud is said to be:
        a. frozen
        b. glaciated
        c. supercooled
        d. supersaturated

ANSWER: b

15.   Which of the following statements is not correct?
        a. Generally, the smaller the pure water droplet, the lower the temperature at
           which it will freeze
        b. Ice nuclei are more plentiful in the atmosphere than condensation nuclei
        c. Much of the rain falling in middle northern latitudes begins as snow
        d. Ice crystals may grow in a cold cloud even though supercooled droplets do not.

ANSWER: b

16.   Ice nuclei may be:
        a. ice crystals
        b. certain clay minerals
        c. bacteria in decaying plant leaf material
        d. all of the above

ANSWER: d

17.   Supercooled cloud droplets are:
        a. ice crystals surrounded by air warmer than 0 °C (32 °F)
        b. liquid droplets that are cooler than the air around them
        c. liquid droplets observed at temperatures below 0 °C (32 °F)
        d. water droplets that have had all their latent heat removed

ANSWER: c

18.   At the same sub-freezing temperature, the saturation vapor pressure just above a liquid water
        surface is _____ the saturation vapor pressure above an ice surface.
        a. greater than
        b. the same as
        c. less than

ANSWER: a

19.   Contact nucleation is:
        a. the freezing of supercooled droplets by contact with a nucleus
        b. the sticking together of ice crystals to make a snowflake
        c. the joining of many nuclei to form an ice nucleus
        d. the freezing of supercooled droplets when the come into contact with a
           supercooled surface

ANSWER: a

20.   The growth of a precipitation particle by the collision of an ice crystal (or snowflake) with a
        supercooled liquid droplet is called:
        a. accretion
        b. spontaneous nucleation
        c. condensation
        d. deposition

ANSWER: a

21.   Cloud seeding using silver iodide only works in:
        a. cold clouds composed entirely of ice crystals
        b. warm clouds composed entirely of water droplets
        c. cold clouds composed of ice crystals and supercooled droplets
        d. cumuliform clouds

ANSWER: c

22. Which of the following conditions would be most suitable for natural cloud seeding by ice crystals?
   a. cirrus clouds above stratus clouds
   b. cirrus clouds above altostratus clouds
   c. altocumulus clouds above stratus clouds
   d. stratus clouds above ground fog
   e. cirrus clouds above cumulus humilis clouds

ANSWER: b

23. What are the two main substances used in cloud seeding?
   a. lead iodide, dry ice
   b. silver iodide, lead iodide
   c. ice crystals, floo powder
   d. dry ice, sea salt
   e. silver iodide, dry ice

ANSWER: e

24. After a rainstorm, visibility typically
   a. deteriorates
   b. is unaffected
   c. improves

ANSWER: c

25. Rain which falls from a cloud but evaporates before reaching the ground is referred to as:
   a. sleet
   b. virga
   c. graupel
   d. dry rain

ANSWER: b

26. The most common ice crystal shape:
   a. graupel
   b. dendrite
   c. rime
   d. virga

ANSWER: b

27.   Fall streaks usually_____ before reaching the ground.
       a. evaporate
       b. condense
       c. sublimate
       d. deposit

ANSWER: c

28.   Snowflakes or ice crystals falling from high cirriform clouds are called:
       a. graupel
       b. snow squalls
       c. snow flurries
       d. rime
       e. fall streaks

ANSWER: e

29.   A light shower of snow that falls intermittently from cumuliform clouds for a short duration is known as:
       a. snow flurries
       b. a snow squall
       c. virga
       d. a cloud burst
       e. a fall streak

ANSWER: a

30.   The creaking sound produced by walking on snow is most common on:
       a. new snow when the air temperature is below -10 °C (14 °F)
       b. old snow when the air temperature is above -10 °C (14 °F)
       c. old snow when the air temperature is near freezing
       d. new snow when the air temperature is near freezing

ANSWER: a

31.   Which below best describes why a fluffy covering of snow is able to protect sensitive plants and their root systems from damaging low temperatures.
       a. snow is a good insulator
       b. melting snow releases latent heat
       c. snow is a good emitter of infrared energy
       d. snow is a good reflector of sunlight

ANSWER: a

156

32.   Large, heavy snowflakes are associated with:
        a. dry air and temperatures well below freezing
        b. moist air and temperatures well below freezing
        c. dry air and temperatures near freezing
        d. moist air and temperatures near freezing

ANSWER: d

33.   Fall streaks most often form with:
        a. nimbostratus clouds
        b. cumulonimbus clouds
        c. stratus clouds
        d. altostratus clouds
        e. cirrus clouds

ANSWER: e

34.   A true blizzard is characterized by:
        a. low temperatures
        b. strong winds
        c. reduced visibility
        d. blowing snow
        e. all of the above

ANSWER: e

35.   The largest snowflakes would probably be observed in ___ air whose temperature is ___ freezing.
        a. moist, near
        b. dry, near
        c. moist, well below
        d. dry, well below

ANSWER: a

36.   In order for falling snowflakes to survive in air with temperatures much above freezing, the air
        must be _____ and the wet bulb temperature must be _____.
        a. unsaturated; at or below freezing
        b. unsaturated; above freezing
        c. saturated; at or below freezing
        d. saturated; above freezing

ANSWER: a

37. In the winter you read in the newspaper that a large section of the Midwest is without power due to downed power lines. Which form of precipitation would most likely produce this situation?
    a. snow
    b. hail
    c. freezing rain
    d. sleet
    e. rain

ANSWER: c

38. Which is not a correct association?
    a. snow grains - hail
    b. ground blizzard - drifting and blowing snow
    c. snow squall - intense snow shower
    d. sleet - ice pellet
    e. freezing rain - glaze

ANSWER: a

39. Which of the following might be mistaken for hail?
    a. virga
    b. graupel
    c. dendrite
    d. supercooled droplet

ANSWER: b

40. A raindrop or partially melted snowflake that freezes into a pellet of ice in a deep subfreezing layer of air near the surface is called:
    a. snow
    b. freezing rain
    c. sleet
    d. hail
    e. a snow pellet

ANSWER: c

41. Which type of precipitation would most likely form when the surface air temperature is slightly below freezing and the air temperature increases as you move upward away from the ground?
    a. freezing rain
    b. hail
    c. rain
    d. snow
    e. drizzle

ANSWER: a

42. The primary method used in preventing the growth of large, destructive hailstones is to inject a thunderstorm with large quantities of:
    a. silver iodide
    b. ice crystals
    c. dry ice
    d. hydrophobic nuclei
    e. hailstone embryos

ANSWER: a

43. Hail deposited in a long narrow band is known as a:
    a. hail line
    b. hail streak
    c. squall line
    d. ice line
    e. hail front

ANSWER: b

44. Precipitation with the greatest size (diameter) is:
    a. the snow pellet
    b. the snow grain
    c. a hailstone
    d. sleet
    e. a rain drop

ANSWER: c

45. Glaze is another name for:
    a. rime ice
    b. snow
    c. sleet
    d. frost
    e. freezing rain

ANSWER: e

46. You would use a wooden stick to measure rainfall in the:
    a. tipping bucket rain gauge
    b. standard rain gauge
    c. weighing rain gauge
    d. Ozarks

ANSWER: b

47. An amount of precipitation measured to be less than one hundredth of an inch (0.25 mm) is called:
   a. a trace
   b. drizzle
   c. light rain
   d. mist

ANSWER: a

48. If a city were to receive 1/2 inch of rain in the morning and then 5 inches of snow that afternoon, about how much precipitation would the weather service report for that day?
   a. 5 1/2 inches
   b. 1/2 inch
   c. 1 inch
   d. 10 inches

ANSWER: c

49. After a snowstorm the newspaper reports that Buffalo, New York received 1.50 inches of precipitation. If we assume an average water equivalent ratio for this snowstorm, then Buffalo received about ___ inches of snow.
   a. 3
   b. 1.5
   c. 10
   d. 9
   e. 15

ANSWER: e

50. On average, the water equivalent of 10 inches of snow is about ___ inches of water.
   a. 0.5
   b. 1
   c. 2
   d. 2.5
   e. 5

ANSWER: b

51. Radar gathers information about precipitation in clouds by measuring the:
   a. energy emitted by the precipitation particles
   b. absorption characteristics of falling precipitation
   c. amount of energy reflected back to a transmitter
   d. amount of sunlight scattered off the precipitation
   e. amount of solar energy passing through the cloud

ANSWER: c

52. In a typical advancing winter storm, which of the following sequences of precipitation types is most likely to occur?
     a. rain, freezing rain, snow, sleet
     b. rain, sleet, freezing rain, snow
     c. freezing rain, rain, sleet, snow
     d. rain, freezing rain, sleet, snow

ANSWER: d

Essay Exam Questions

1.   Is silver iodide used as a cloud seeding agent in warm or cold clouds? Why?

2.   Explain why it is possible for an ice crystal to grow in a cold cloud even though the supercooled water droplets surrounding the ice crystal do not.

3.   Would you expect the largest forms of precipitation particles to occur during the warmest or the coldest time of year? Explain.

4.   What is the main difference between a raindrop and a cloud droplet?

5.   The first raindrops to reach the ground at the beginning of a rain shower are often very large. Explain why this is so.

6.   Thunderstorm cloud bases are generally higher above the ground in Arizona than in Florida. Why?

7.   Briefly describe the differences between snow, freezing rain, sleet and hail.

8.   About how large can raindrops get? Why can't they get any larger? What shape does a large raindrop have? What forces determine this shape?

9.   Would you expect the heaviest snowfall to occur on an unusually cold night or a night when the temperature was just a little below freezing?

10. Explain why it is much more difficult to measure snowfall amount than rainfall amount.

11. Design an automated device for measuring snowfall. How might it work?

Chapter 9
The Atmosphere in Motion:
Air Pressure, Forces, and Winds

Summary

This chapter provides a broad view of how and why the wind blows. The chapter begins by reviewing and extending some of the basic concepts about pressure introduced earlier in the text. These concepts include the ideal gas law, which is described in a lengthy focus section. Horizontal temperature variations are shown, for example, to produce pressure gradients that can cause the air to move. Instruments used to measure pressure and the most common pressure units are discussed. Examples of meteorological charts used to display surface and upper level pressure patterns are presented.

Newton's laws of motion are defined and forces that govern the horizontal movements of air are identified. The relatively simple case of air motions above the ground is studied first. Winds aloft are affected by just the pressure gradient force and the Coriolis force. Horizontal pressure gradients initially set the air in motion; the Coriolis force then exerts a force to the right or left of the wind's direction of motion. Winds at upper levels blow parallel to the contour lines on an isobaric chart. When the contours lines are straight, the Coriolis and pressure gradient forces are equal and opposite and the wind blows in a straight line at constant speed. When the flow is curved, the resulting gradient wind forces include a centripetal component which accounts for the changes in wind direction. Some details of these relationships between the different forces are elaborated in focus sections on geostrophic wind, isobaric (constant pressure) surfaces and the hydrostatic equation.

The frictional force acts to slow wind speeds, at the ground, with the result that winds blow across the isobars slightly toward lower pressure. This accounts for the rising and sinking air motions found in high and low pressure centers. Converging or diverging air motions above the ground can cause surface features to strengthen or weaken.

Key Terms

air pressure
(atmospheric) model
gas law
equation of state
net convergence
net divergence
thermal tides
barometer
millibar
hectopascal
inches of mercury (Hg)
standard atmospheric
  pressure
Pascal
barometric pressure
mercury barometer
aneroid barometer
altimeter
barograph
instrument error
station pressure

altitude correction
mean sea level
sea level pressure
isobars
constant height chart
constant pressure
  (isobaric) chart
isobaric surface
contour lines (on
  isobaric charts)
ridges
troughs
anticyclones
depressions
mid-latitude cyclones
contour interval
Newton's first law
  of motion
Newton's second law
  of motion
pressure gradient

pressure gradient
  force (PGF)
Coriolis force
apparent force
Coriolis effect
geostrophic wind
Coriolis parameter
gradient wind
centripetal acceleration
centripetal force
cyclostrophic
centrifugal force
gradient wind equation
zonal flow
meridional flow
friction layer
Buys-Ballot's law
hydrostatic equilibrium
hydrostatic equation

Teaching Suggestions

1.   A one-inch-square iron bar cut approximately 53 inches long weighs 14.7 pounds. When placed on end, the pressure at its base will be 14.7 pounds/sq. in., the same as that of the atmosphere at sea level. The bar can be passed around class and the students will be surprised at how heavy it is. The bar can be used to motivate a discussion of the concept of density and the workings of the mercury barometer. If the bar were constructed of mercury, it would only need to be 30 inches long. If the bar were made of water, it would need to be close to 34 feet tall. Students should understand also why they are not crushed by the weight of many "iron bars" pressing in on every square inch of their bodies. One misconception that the bar might create is that pressure exerts only a downward force. The next demonstration might help clear that up.

2.   There are a variety of "crushed-can" demonstrations. For example, put a small amount of water into a clean, metal can. Heat the can until the water boils and then tightly seal the spout. The can will be crushed by the weight of the atmosphere as it cools. The water is not necessary, but it enhances the effect. Use the ideal gas law to explain the pressure imbalance that was created.

3.   Show students a surface weather map without any isobars drawn on it. The students will appreciate how difficult it is to assimilate the large quantity of data plotted on the map. It will not be apparent at all what large scale weather features are present nor what is causing the observed weather conditions. Then, show the students the same map with a completed isobaric analysis. The positions of important high and low pressure centers will become clear immediately and their effect on the weather

conditions in their vicinity will be apparent.

4.   Any discussion of the Coriolis force will lead to a question about whether the direction water spins when draining out of a sink or toilet bowl is different in the Southern Hemisphere than it is in the Northern Hemisphere. The instructor can demonstrate that water can be made to rotate in either direction when draining from a plastic soft drink bottle.

5.   Demonstrate the difference between pressure and force by having a student lean on another student using his/her open palm, then repeating the process using only a fingertip. Same force, different pressures.

Blue Skies 6.   Demonstrate the Coriolis force using the Atmospheric Forces/Coriolis Force flight simulator activity on the BlueSkies cdrom.

Student Projects

1.   Give the students a simplified surface weather chart and have them perform an isobaric analysis. Initially, to keep the map as simple as possible, it might be best to plot only the wind and pressure data. Once students have located centers of high and low pressure, have them transfer the positions to a second map with additional data (temperature, cloud cover, weather, dew point). Have students circle regions with overcast skies or stations that are reporting precipitation. Are the stormy regions associated with high or low pressure?

2.   Have students plot daily average pressure and observed weather condition for several weeks. Are stormy periods well correlated with lower-than-average pressures?

3.   Determine the direction of upper-level winds by observing mid-level cloud motions. Have students draw the orientation of contour lines that would produce the observed motion and then compare their sketch with an actual upper-air chart.

4.   Have students obtain examples of weather reports and weather maps from other countries from the web, or from the foreign newspaper collection at the university or local library. Students should attempt to find at least one example from a city in the Southern Hemisphere.

Blue Skies 5.   Using the Atmospheric Forces/Winds in Two Hemispheres activity on the BlueSkies cdrom, count the numbers of Highs and Lows with at least two closed isobars that may be found currently anywhere on the earth. Are most of the highs and lows close to the equator? Why or why not?

Blue Skies 6.   Using the Atmospheric Forces/Winds in Two Hemispheres activity on the BlueSkies cdrom, determine whether Buys Ballot's Law has to be modified for application to the Southern Hemisphere.

Answers to Questions for Thought

1.   Assuming no corrections are made for altitude, the drop in pressure from the base to the top of the hill would be detected by the barometer; the reading could be lower than that observed in most storms.

2.   Inside the refrigerator, the temperature of the air inside the balloon will decrease. The Gas Law requires that the air pressure inside the balloon must also decrease. (The air density inside the balloon will also increase slightly.)

3.   Air column T will have the highest surface pressure (dry air weighs more than moist air at the same temperature).

4.   Sure, if the station was located *below* sea-level. (One example would be Death Valley, California.)

5.   Higher pressures aloft are to the northwest, lower pressures are to the southeast. Since there is a significant north-south component to the upper-level air flow, the flow is meridional.

6.   High pressures and high temperatures are associated with high heights on an isobaric surface, while low pressures and low temperatures are associated with low heights. When flying from high pressure and warm air into a region of low pressure and cold air, without changing the altimeter setting, the altimeter (which measures pressure and indicates altitude) will indicate an altitude higher than the aircraft is flying. Because the aircraft is flying lower than indicated, it is wise to look out below for obstructions such as mountains and hills.

7.   The altimeter will continue to indicate the same altitude because it measures pressure. The constant pressure surface will be located at higher than standard altitude in the warmer air because pressure decreases more slowly in warm air than in cooler air. Once it enters the warm air, the aircraft will actually be flying at a higher altitude than that indicated by the altimeter.

8.   The wind would blow directly from regions of higher pressure toward regions of lower pressure.

9.   The frictional effect of the water is less. Consequently, with the same PGF the winds are stronger, the Coriolis force is stronger, and the winds blow more nearly parallel to the isobars.

10.   Assuming that this flow is in the Northern Hemisphere, the flow would be cyclonic (low pressure center). The relative magnitude of the centripetal force would be the difference between the PGF and Coriolis force. The centripetal force would be directed inward.

12.   The surface wind would probably be either south or southeast. If the upper-level low is an elongated trough of low pressure, then the winds overhead would probably be more southwesterly, and the middle-level clouds would be moving from southwest to northeast. The wind is changing in a clockwise direction with increasing height, this is veering. The winds aloft would be stronger than the surface winds.

13.   By illustrating the wind flow patterns around a high pressure area and moving the high from west to east, north of your location.

14.   On a frictionless surface the winds around a surface low would blow parallel to curved isobars with the PGF directed inward and the Coriolis force directed outward. However, frictions slows the wind, which reduces the Coriolis force. Because of the stronger PGF, the winds blow inward across the isobars toward lower pressure, giving the appearance that the winds are being deflected to the left. The inwardly-directed force is stronger.

15.   On the equator the Coriolis force is zero. Because there is no deflecting force the wind can blow either clockwise or counterclockwise around an area of low pressure, depending upon how the flow initially responds to the PGF. In either case, the PGF supplies the inwardly-directed force needed for circular motion. This would be an example of cyclostrophic flow.

16.   Inside the closed car the air density remains constant, however, the temperature increases. In this case the gas law ($P = T$ x constant) dictates that an increase in temperature will produce an increase in pressure. If the car is air tight, the pressure inside could become great enough to crack or "blow out" a window.

17.   The wind speed will increase due to reduced friction over the lake. The wind direction will veer (turn in a clockwise direction) as a result of the increased Coriolis force.

18.   The magnitude of the Coriolis force is very small close to the equator. Also, the slope of the tub basin is a much more important factor in determining the direction of rotation.

Answers to Problems and Exercises

1.   Assuming standard atmospheric conditions, the sea-level pressure would be 1024 mb. On a hot afternoon, the actual sea-level pressure would be somewhat less.

2.   (a) 997 mb, 999 mb
       (c) At Point A probably southeasterly, at Point B probably westerly
       (d) Stronger winds at Point B, greater PGF
       (e) 8 mb/1000 km between Points 1 and 2, and 12 mb/1000 km between Points
          3 and 4
       (f) Geostrophic wind at Point A 9.13 m/sec (about 18 knots) geostrophic wind at
          Point B 13.7 m/sec (about 27 knots)
       (g) Due to friction, the winds at Points A and B would be less than the
          geostrophic wind computed in part (f).

3.   (a) Pressure at top of column about 500 mb as $\Delta P = 49940.8$ N/m^2 or
          P(at top) = 1000 - 499.4 = 500.6 mb.
       (b) Pressure would be less than that computed in part (a) because the air density
          would increase as the column cooled and air pressure decreases more rapidly
          with height in cold, high density air.
       (c) If we assume the column is still 5600 m thick, and the atmospheric pressure
          at the bottom of the column is 1000 mb, then the pressure at the top would be
          468 mb as P = 532 mb and P(at top) = 1000 - 532 = 468 mb.

4.   P = 502.25 mb; observed at about 5.5 km (18,000 feet).

5.   (a) density = 1.19 kg/m^3
        (b) Within the container, an increase in pressure must be caused by an increase
            in temperature. New temperature = 586 K.

6.   P = 1009 mb.

7.   To the east.

Multiple Choice Exam Questions

1.   Net convergence of air would cause surface pressure to ____ and net divergence would cause
        surface pressure to __.
        a. increase, decrease
        b. increase, increase
        c. decrease, decrease
        d. decrease, increase

ANSWER: a

2.   If the earth's gravitational force were to increase, atmospheric pressure at the ground would:
        a. increase
        b. decrease
        c. remain the same
        d. cause the atmosphere to expand vertically

ANSWER: a

3.   The surface pressures at the bases of warm and cold columns of air are equal. Which of the
        following statements is not correct?
        a. pressure will decrease with increasing height at the same rate in both columns
        b. the cold air is more dense than the warm air
        c. both columns of air contain the same total number of air molecules
        d. the weight of each column of air is the same

ANSWER: a

4.   The surface pressures at the bases of warm and cold columns of air are equal. Air pressure in the
        warm column of air will __ with increasing height __ than in the cold column.
        a. decrease, more rapidly
        b. decrease, more slowly
        c. increase, more rapidly
        d. increase, more slowly

ANSWER: b

5.   Suppose a parcel of air has a given temperature, pressure, and density. If the parcel's size remains the same while its temperature increases, then the air pressure inside the parcel will:
   a. decrease
   b. decrease to but not lower than 1000 mb
   c. increase
   d. remain constant

ANSWER: c

6.   Suppose a very cold parcel of air at 5.5 km (18,000 feet) is compared to a similar (but warm) parcel of air at sea level. Which of the following would be true?
   a. the parcel at sea level has higher pressure and higher density
   b. the parcel at sea level has lower pressure and higher density
   c. the parcel at sea level has lower pressure and lower density
   d. the parcel at sea level has higher pressure and lower density

ANSWER: a

7.   If two air parcels at sea level have the same size but different temperatures, the colder parcel of air will have:
   a. a higher pressure but the same density as the warm parcel
   b. the same pressure but lower density than the warm parcel
   c. the same pressure but higher density than the warm parcel
   d. a lower pressure but the same density as the warm parcel

ANSWER: c

8.   Which of the following relationships best describes the gas law?
   a. pressure is proportional to density times temperature
   b. density is proportional to pressure times temperature
   c. temperature is proportional to density times pressure
   d. temperature times pressure times density remains constant

ANSWER: a

9.   If surface air pressure decreases, the height of the column in a mercury barometer would:
   a. remain constant
   b. increase
   c. decrease
   d. change momentarily, but return to its earlier reading

ANSWER: c

10.   If the outside air temperature is 27 °C and the air density is 1.2 kg/m³, the outside air pressure
      would be:
      a. 32 mb
      b. 93 mb
      c. 930 mb
      d. 1013 mb
      e. 1033 mb

ANSWER: d

11.   The scale on an altimeter indicates altitude, but an altimeter actually measures:
      a. temperature
      b. density
      c. pressure
      d. humidity

ANSWER: c

12.   A barograph will:
      a. measure and record atmospheric pressure
      b. maintain the pressure in a room at a preset value
      c. predict pressures 1 or 2 days before they occur
      d. draw isobars on a surface weather map

ANSWER: a

13.   The unit of pressure most commonly found on a surface weather map:
      a. inches of mercury
      b. millibars
      c. pounds per square inch
      d. atmospheres

ANSWER: b

14.   If a liquid with a lower density than mercury were used in a barometer the height of the column in
      the barometer would:
      a. increase
      b. decrease
      c. remain the same
      d. not provide an accurate measure of atmospheric pressure

ANSWER: a

15.   An aneroid barometer works on the principle that:
       a. mercury will rise and descend in a tube when the air pressure changes
       b. the force of gravity decreases in strength with increasing altitude
       c. a small closed cell with most of its air removed will expand and contract with
          changes in air pressure
       d. a change in pressure causes a weak electrical signal in a ceramic detector
       e. a change in air pressure causes a simultaneous change in air temperature that
          is detected with a sensitive thermometer

ANSWER: c

16.   Which of the following instruments measures pressure?
       a. barometer
       b. thermometer
       c. radiometer
       d. hygrometer
       e. densitometer

ANSWER: a

17.   An aneroid barometer carried from sea level to the top of a 300 m (1000 ft) hill would **indicate**:
       a. low humidity
       b. stormy weather
       c. clear skies
       d. fair weather

ANSWER: b

18.   The mercury barometer was invented by:
       a. Newton
       b. Coriolis
       c. Torricelli
       d. Aristotle

ANSWER: c

19.   To obtain the <u>station pressure</u> you must normally make corrections for:
       a. temperature and gravity
       b. temperature and altitude
       c. operator error and temperature
       d. temperature and time

ANSWER: a

20.   A station at an altitude of 900 m (about 3000 feet) above sea level measures an air pressure of 930 mb. Under normal conditions, which of the values below do you think would be the most realistic sea level pressure for this station?

a. 840 mb
b. 930 mb
c. 1020 mb
d. 1830 mb

ANSWER: c

21.   Suppose a station at sea level measures an air pressure of 1030 mb. Under standard conditions, what would be the most likely air pressure at an elevation of 600 m (about 2000 ft) above this station?
a. 1090 mb
b. 1030 mb
c. 1024 mb
d. 1010 mb
e. 970 mb

ANSWER: e

22.   To correctly monitor horizontal changes in air pressure, the most important correction for a mercury barometer is the correction for:
a. temperature
b. altitude
c. density
d. gravity

ANSWER: b

23.   The surface weather map is a sea level chart. Thus, a surface weather map is also called:
a. a constant pressure chart
b. a constant height chart
c. an isobaric chart
d. a constant latitude chart

ANSWER: b

24.   Lines connecting points of equal pressure are called:
a. isobars
b. millibars
c. contours
d. isotherms
e. a coordinate grid

ANSWER: a

25.   Pressure changes:
        a. more rapidly in the horizontal direction than in the vertical
        b. more rapidly in the vertical direction than in the horizontal
        c. at the same rate in the horizontal and vertical directions
        d. more rapidly in the vertical over land than over the ocean.

ANSWER: b

26.   On a weather map, ridges are:
        a. elongated low pressure areas
        b. dying hurricanes
        c. mountains that stall the movement of storms
        d. elongated high pressure areas
        e. tornadoes that touch the surface

ANSWER: d

27.   The gas law is
        a. $p \propto T \times \rho$
        b. $T \propto p \times \rho$
        c. $\rho \propto T \times p$

ANSWER: a

28.   According to the gas law, when temperature remains constant,
        a. $\rho \propto p$
        b. $\rho \propto 1/p$
        c. $p \propto 1/\rho$
        d. both b and c
        e. none of the above

ANSWER: a

29.   Suppose you are a pilot who is flying from warm air into colder air. In the cold air, even though
        your altimeter is still indicating the same altitude as it did in the warm air, you would be flying:
        a. higher than you were before
        b. faster than you were before
        c. slower than you were before
        d. at the same altitude as you were before
        e. lower than you were before

ANSWER: d

30. On an isobaric surface,
   a. altitude is constant
   b. temperature is constant
   c. pressure is constant
   d. both a and c

ANSWER: c

31. Low ___ on a constant height chart corresponds to low ___ on a constant pressure chart.
   a. pressures, pressures
   b. pressures, heights
   c. heights, pressures
   d. heights, heights

ANSWER: b

32. On a 500 millibar chart, _____ are drawn to represent horizontal changes in altitude which correspond to horizontal changes in pressure.
   a. contour lines
   b. isobars
   c. isotherms
   d. isotachs

ANSWER: a

33. The contour lines drawn on a 500 mb chart are lines of constant:
   a. pressure
   b. altitude
   c. density
   d. wind direction

ANSWER: b

34. If an airplane flies from standard temperature air into warmer than standard temperature air, without making any correction, the altimeter in the warmer air would indicate an altitude:
   a. higher than the airplane's actual altitude
   b. exactly the same as the airplane's altitude
   c. lower than the airplane's actual altitude
   d. correction factor to be used by the pilot

ANSWER: c

35. Warm air aloft is associated with constant pressure surfaces that are found at ___ altitude than normal and ___ than normal atmospheric pressure aloft.
    a. higher, higher
    b. higher, lower
    c. lower, higher
    d. lower, lower

ANSWER: a

36. On an upper-level chart, normally we find warm air associated with ___ pressure, and cold air associated with ___ pressure.
    a. high, high
    b. high, low
    c. low, low
    d. low, high

ANSWER: b

37. A surface low pressure center is generally associated with ___ on an upper level isobaric chart.
    a. a trough
    b. a ridge
    c. zonal flow
    d. convergence

ANSWER: a

38. On an upper-level chart the wind tends to blow:
    a. at right angles to the isobars or contour lines
    b. parallel to the isobars or contours
    c. at an angle between 10 and 30 to the contours and towards lower pressure
    d. at constant speed

ANSWER: b

39. A ridge on an upper-level isobaric chart indicates:
    a. higher-than-average heights
    b. lower-than-average heights
    c. average heights
    d. a region with calm winds

ANSWER: a

40. On an isobaric weather chart, the spacing of the height contours indicates the magnitude of the pressure gradient force.
    a. true
    b. false

ANSWER: a

41. Newton's first law states that *an object at rest* _____ *as long as no force is exerted on the object.*
    a. *equals its mass times the acceleration produced*
    b. *will remain at rest*
    c. *can exert no force*

ANSWER: b

42. During a reversible adiabatic process, the pressure gradient force is parallel to the isobars.
    a. true
    b. false

ANSWER: b

43. The Coriolis force is the force that causes the wind to blow.
    a. true
    b. false

ANSWER: b

44. Given the equation for geostrophic wind, $V_g = \dfrac{1}{2\Omega\sin\phi\rho}\dfrac{\Delta p}{d}$, increasing the pressure gradient ($\dfrac{\Delta p}{d}$) will increase the wind speed.
    a. true
    b. false

ANSWER: a

45. Given the equation for geostrophic wind, $V_g = \dfrac{1}{2\Omega\sin\phi\rho}\dfrac{\Delta p}{d}$, increasing the angular spin of the earth (Ω) will increase the wind speed.
    a. true
    b. false

ANSWER: b

46. The hydrostatic equation describes the equilibrium between the
    a. friction force and gravity
    b. horizontal pressure gradient force and gravity
    c. Coriolis force and gravity
    d. horizontal pressure gradient force and gravity
    e. vertical pressure gradient force and gravity

ANSWER: e

47.   Why is there a minus sign in the hydrostatic equation?   (The hydrostatic equation is $\Delta p = -\rho g \Delta z$.)
        a. Because as pressure decreases, gravity increases.
        b. Because as pressure decreases, density increases.
        c. Because as pressure decreases, height increases.
        d. Because as pressure decreases, height decreases.

ANSWER: c

48.   The fundamental laws of motion were formulated by:
        a. Galileo
        b. Newton
        c. Coriolis
        d. Aristotle
        e. Plato

ANSWER: b

49.   Newton would have been alive in:
        a. 1492
        b. 1647
        c. 1835
        d. 1934
        e. 1066

ANSWER: b

50.   The "force exerted on an object equals its mass times the acceleration produced" is a description of:
        a. Newton's second law of motion
        b. Buys-Ballot's law
        c. the isobaric law
        d. hydrostatic equilibrium

ANSWER: a

51.   Which of the following forces does not have a direct effect on horizontal wind motions?
        a. pressure gradient force
        b. frictional force
        c. gravitational force
        d. Coriolis force

ANSWER: c

52.   Which of the following can influence wind direction?
       a. Coriolis force
       b. pressure gradient force
       c. centripetal force
       d. all of the above

ANSWER: d

53.   Which of the following forces can <u>not</u> act to change the speed of the wind?
       a. pressure gradient force
       b. frictional force
       c. Coriolis force
       d. none of the above

ANSWER: c

54.   The net force on air moving in a circle at constant speed is:
       a. inward toward the center of rotation
       b. zero
       c. in the direction of wind motion
       d. outward from the center of rotation

ANSWER: a

55.   The amount of pressure change that occurs over a given horizontal distance is called the:
       a. pressure tendency
       b. Coriolis parameter
       c. pressure gradient
       d. potential gradient
       e. slope

ANSWER: c

56.   Which of the statements below is <u>not</u> correct concerning the pressure gradient force?
       a. the PGF points from high to low pressure in the Northern Hemisphere
       b. it is non-existent at the equator
       c. it can cause the wind to speed up or slow down
       d. the PGF points from high to low pressure in the Southern Hemisphere

ANSWER: b

57.   The pressure gradient force is directed from higher pressure toward lower pressure:
       a. only at the equator
       b. at all places on earth except for the equator
       c. only in the Northern Hemisphere
       d. only in the Southern Hemisphere
       e. at all places on earth

ANSWER: e

58.   The force that would cause a stationary parcel of air to begin to move horizontally is called the:
       a. Coriolis force
       b. pressure gradient force
       c. centripetal force
       d. frictional force

ANSWER: b

59.   Which of the following produces the strongest Coriolis force?
       a. fast winds, low latitude
       b. fast winds, high latitude
       c. slow winds, low latitude
       d. slow winds, high latitude

ANSWER: b

60.   The ___ is an apparent force created by the earth's rotation.
       a. pressure gradient force
       b. Coriolis force
       c. centripetal force
       d. gravitational force

ANSWER: b

61.   The rate of the earth's rotation determines the strength of the:
       a. pressure gradient force
       b. Coriolis force
       c. frictional force
       d. gravitational force

ANSWER: b

62.   The Coriolis force is the result of:
       a. wind
       b. rotating earth
       c. day/night temperature contrasts
       d. gravitational force exerted by the moon as it orbits the earth
       e. the poles being colder than the equator

ANSWER: b

63.   Which statement below is <u>not</u> correct concerning the Coriolis force?
       a. It causes the winds to deflect to the right in the Northern Hemisphere
       b. It is strongest at the equator
       c. It can cause winds to change direction, but not to increase or decrease in speed
       d. It deflects winds in opposite directions in the Northern and Southern
          Hemispheres.

ANSWER: b

64.   If the earth stopped rotating which of the following would <u>not</u> be true?
       a. surface winds would below from high toward low pressure
       b. there would still be a Coriolis force
       c. there would still be a pressure gradient force
       d. there would still be a gravitational force

ANSWER: b

65.   A wind blowing at a constant speed parallel to straight line isobars with the pressure gradient
       force (PGF) and the Coriolis force in balance is called a:
       a. gradient wind
       b. meridional wind
       c. geostrophic wind
       d. cyclostrophic wind
       e. zonal wind

ANSWER: c

66.   The net force acting on air which is blowing parallel to straight contours at constant speed is:
       a. in the direction of wind motion
       b. to the right of the wind's motion in the Northern Hemisphere
       c. zero
       d. in a direction opposite the wind's motion

ANSWER: c

67. Suppose that the winds aloft are geostrophic and blowing from the north. Low pressure is located to the:
   a. north
   b. south
   c. east
   d. west

ANSWER: c

68. Suppose that the winds aloft are geostrophic and blowing from the north. With the same orientation of isobars at the surface, the winds would blow from the:
   a. southwest
   b. northwest
   c. northeast
   d. southeast

ANSWER: b

69. The winds aloft in the middle latitudes would not blow from the west if:
   a. the earth's rotation slowed or increased slightly
   b. the tilt of the earth changed slightly
   c. the air over high latitudes became warmer than over the equator
   d. the direction of the moon's orbit around the earth were reversed

ANSWER: c

70. If in the Northern Hemisphere the upper level winds above you are blowing from the south, then it is a good bet that a trough of low pressure is to the ___ of you.
   a. north
   b. south
   c. east
   d. west

ANSWER: d

71. When the wind blows in a more or less west to east direction, the wind flow pattern is called:
   a. gradient
   b. meridional
   c. centripetal
   d. zonal

ANSWER: d

72. A wind that blows at a constant speed parallel to curved isobars or contour lines is called a:
    a. geostrophic wind
    b. cyclonic wind
    c. convergent wind
    d. gradient wind

ANSWER: d

73. If directly above you at 10,000 feet the wind is blowing from the south, then it is a good bet that at 10,000 feet, the center of <u>lowest</u> pressure is ___ of you, while the center of <u>highest</u> pressure is _ of you.
    a. west, east
    b. south, north
    c. east, west
    d. north, south

ANSWER: a

74. A wind flow pattern that takes on a more or less north-south trajectory is called:
    a. gradient flow
    b. zonal flow
    c. cyclostrophic flow
    d. meridional flow
    e. geostrophic flow

ANSWER: d

75. The vertical pressure gradient force is directed
    a. downward
    b. upward
    c. horizontally

ANSWER: b

76. The winds aloft in the middle latitudes of the Southern Hemisphere generally blow:
    a. from west to east
    b. from east to west
    c. from north to south
    d. from south to north

ANSWER: a

77.   A surface LOW pressure area that moves from south to north directly east of your home would most likely produce winds that shift from:
        a. S to SE to E
        b. SE to E to SW
        c. N to NE to E
        d. W to NW to N
        e. NE to N to NW

ANSWER: e

78.   The wind around a surface low pressure center in the Southern Hemisphere blows:
        a. counterclockwise and outward from the center
        b. counterclockwise and inward toward the center
        c. clockwise and outward from the center
        d. clockwise and inward toward the center

ANSWER: d

79.   Surface winds blow across the isobars at an angle due to:
        a. the Coriolis force
        b. the pressure gradient force
        c. the frictional force
        d. the centripetal force

ANSWER: c

80.   If, at your home in the Northern Hemisphere, the surface wind is blowing from the northwest, then the region of lowest pressure will be to the ___ of your home.
        a. north
        b. south
        c. east
        d. west

ANSWER: c

81.   Winds blow slightly inward:
        a. around surface low pressure centers in the Northern Hemisphere only
        b. around surface low pressure centers in the Southern Hemisphere only
        c. around surface low pressure centers in the Northern and Southern Hemispheres.
        d. at the poles in both hemispheres

ANSWER: c

182

82.	Cyclonic flow means ___ in either the Northern or Southern Hemisphere.
	a. clockwise wind flow
	b. counterclockwise flow
	c. circulation around a low pressure center
	d. circulation around a high pressure center

ANSWER: c

83.	Buys-Ballot's law states that, "In the Northern Hemisphere if you stand with your back to the surface wind, then turn clockwise about 30°, lower pressure will
	be ___ ."
	a. to your right
	b. to your left
	c. behind you
	d. in front of you

ANSWER: b

84.	The wind around a surface high pressure center in the Northern Hemisphere blows:
	a. counterclockwise and outward from the center
	b. counterclockwise and inward toward the center
	c. clockwise and outward from the center
	d. clockwise and inward toward the center

ANSWER: c

85.	We can generally expect the air to be ___ above areas of surface low pressure and ___ above areas of surface high pressure.
	a. rising, rising
	b. rising, sinking
	c. sinking, sinking
	d. sinking, rising

ANSWER: b

86.	The surface air around a strengthening low pressure area normally ___ , while aloft, above the system, the air normally ___ .
	a. diverges, diverges
	b. diverges, converges
	c. converges, converges
	d. converges, diverges

ANSWER: d

87. The atmosphere around the earth would rush off into space if the vertical pressure gradient force were not balanced by:
    a. the Coriolis force
    b. the horizontal pressure gradient force
    c. gravity
    d. the centripetal force
    e. friction

ANSWER: c

88. In the vertical, the pressure gradient force points ___ and gravity points ___ .
    a. toward the earth, away from the earth
    b. toward the earth, toward the earth
    c. away from the earth, away from the earth
    d. away from the earth, toward the earth

ANSWER: d

89. When the upward-directed pressure gradient force is in balance with the downward pull of gravity, the atmosphere is in:
    a. hydrostatic equilibrium
    b. unstable equilibrium
    c. geostrophic balance
    d. isobaric balance

ANSWER: a

90. If an air parcel is completely at rest, which of the following forces can make the parcel begin to move?
    a. friction
    b. centripetal
    c. pressure gradient
    d. gravity
    e. Coriolis

ANSWER: c

Essay Exam Questions

1. Suppose you stand outside and feel a fresh breeze blowing against your face. Could this be a geostrophic wind? Explain.

2. Under what conditions (if any) might you record a station pressure of 750 mb?

3. What differences might you expect to see between the weather conditions depicted on the surface in your city and at the 500 mb level above your city?

184

4. Suppose you deflate a bicycle tire by depressing the air valve. Why does air rush out of the tire? After the air stops rushing out of the tire, is the tire empty? Explain your answer.

5. Briefly explain the principle of the mercury barometer. Mercury is relatively expensive and toxic. Why do you think mercury is used in barometers instead of another fluid such as water?

6. If the earth were to begin rotating in the other direction, would air still rise in the center of surface low pressure?

7. Is a force needed to keep a satellite orbiting at constant speed around the earth?

8. Sketch the wind flow patterns around surface high and low pressure centers in the Northern and Southern Hemispheres.

9. The pressure announced on last night's television weather broadcast was 29.92. Explain how this was measured and give the units. Would this be considered an unusually large or low pressure value?

10. Explain briefly why upper-level winds at middle latitudes in the Northern Hemisphere blow from West to East. In what direction do upper level winds at middle latitudes in the Southern Hemisphere blow?

11. If the earth did not rotate, how would you expect winds to blow with respect to high and low pressure centers?

12. Explain why closely-spaced contour lines on an upper-level isobaric chart are associated with fast winds.

13. Draw a simple Northern Hemisphere upper-air pressure pattern consisting of several straight, uniformly-spaced contour lines running from left to right across your paper. Assume that lower heights are found at the top of your chart. Use arrows to indicate the direction that the wind would blow and the direction of the pressure gradient force and Coriolis force acting on a moving parcel of air.

14. Explain why strong upper-level divergence will cause the pressure in the center of a surface low to decrease.

15. Explain why it is often windy at the beach. What forces are responsible, and how do beachfront conditions differ from conditions farther inland?

Chapter 10
Wind: Small-Scale and Local Systems

Summary

A wide variety of types of small- and middle-scale wind motions are examined in this chapter, ranging from short-lived microscale phenomena to the seasonal Asian monsoon system. The chapter begins with a quick classification of the scales of atmospheric motion and looks at the formation of eddies. Wind motions can have a variety of effects on the environment, such as lifting and transporting soil, shaping sand dunes, and creating waves in the ocean. Wind shear and turbulent eddies in clear air can present a hazard to aviation.

The formation of thermal circulations is then covered in some detail. Sea, lake, and land breezes are presented as examples of thermal circulations, and the effect that they can have on their surroundings is examined. The Asian monsoon, which, in many ways, resembles a large scale thermal circulation, is covered. Additional local scale wind systems including mountain and valley breezes, katabatic winds, chinook and the Santa Ana winds are discussed, as is the relationship between sea breezes and widespread fires in Florida. An interesting focus section describes a wind circulation on a different planet - dust storms on Mars.

The chapter includes descriptions of some of the methods used to measure wind speed and direction at the ground and upper levels. Examples of some of the practical uses for wind data are given.

Key Terms

wind
scales of motion
microscale
mesoscale
synoptic scale
planetary scale
macroscale
friction
viscosity
molecular viscosity
laminar flow
eddy viscosity
wind gusts
mechanical turbulence
planetary boundary
  layer
thermal turbulence
eddy
roll eddies
rotors
wind shear
clear air turbulence
  (CAT)
air pockets
ridge soaring
saltation
desert pavement
ventifacts
sand dunes
sand ripples
snow ripples
snow dunes
snow rollers
wind sculptured trees
shelterbelts (windbreaks)

wind waves
fetch
swells
onshore wind
offshore wind
upslope wind
downslope wind
prevailing wind
wind rose
wind vane
wind sock
anemometer
pressure plate
  anemometer
cup anemometer
aerovane (skyvane)
pilot balloon
theodolite
radiosonde
rawinsonde
lidar
wind sounding
wind profiler
wind turbines
thermal circulations
thermal highs
thermal lows
sea breeze
lake breeze
land breeze
sea breeze front
smoke (smog) front
sea breeze
  convergence zone

standing waves
seiches
monsoon wind system
monsoon depression
Southern Oscillation
Pacific monsoon
valley breeze
mountain breeze
gravity winds
drainage winds
katabatic (or fall)
  wind
bora
mistral
Columbia Gorge wind
chinook wind
foehn wind
compressional heating
chinook wall cloud
Santa Ana wind
California norther
sandstorms
haboob
dust devils
  (whirlwinds)
leste
leveche
sirocco
khamsin
sharav
Texas norther
northeasters
Boulder winds
mountainadoes

Teaching Suggestions

1.   Clouds produced by some of the mesoscale wind circulation patterns discussed in this chapter can be seen on satellite photographs. Sea breeze convergence zones are often visible along the eastern or western coastline of Florida. Ask students why a strong convergence zone will form along the west coast one day and then along the east coast another day.

2.   Construct a box approximately 12" wide x 12" high x 4" deep. The front side of the box should be constructed out of clear plastic. Place two "cat food" cans at opposite ends of the bottom of the box. Fill one can with dark soil, the other with water. Place a 150 or 200 Watt bulb so that light shines equally onto the two cans in the box. After a few minutes, carefully introduce some smoke into the chamber from a hole near the middle of the bottom edge of the plastic. With some care, a circular circulation will be visible.

Student Projects

1.   Have students research and summarize the weather conditions that produce any local scale winds that are unique to their area. Do these local winds have any important effects on the local climate?

2.   Have students determine the prevailing winds and construct a windrose for their city or town. Can students find any evidence that data of this type has been used by city planners? (Exercise 1 at the end of the chapter in the text list some of the ways prevailing wind data might be used in city design)

3.   Have students summarize the weather conditions prevailing during a period when strong Santa Ana winds are being observed in southern California.

4.   Have students plot hourly wind data for their city for a period of several days. At what time of day do the strongest winds occur? Have students discuss and interpret their findings.

Blue Skies 5.   Use the Weather Analysis/Find the Front section of the BlueSkies cdrom to examine temperature and cloud cover over Florida. Discuss whether conditions are favorable for the development of a sea breeze circulation.

Blue Skies 6.   Using the Atmospheric Forces/Winds in Two Hemispheres activity on the BlueSkies cdrom, examine the 24-hour history of a U.S. weather map with wind vector overlays. Can you find any evidence of small-scale circulations, such as thermal circulations or sea/land breezes.

Answers to Questions for Thought

1.   Probably in the early morning before sunrise when the atmosphere is most stable and thermal turbulence is at a minimum.

2.   Turbulent eddies form on the leeward side of a hill during periods of strong winds. The downward motion of the eddies can cause a hang-glider to crash into the ground or along the side of a hill.

188

3.   North, south, and east of Cheyenne, Wyoming, are fairly flat plains (mountains lie to the west). High winds pick up and carry some of the snow deposited on the plains. A town on the plains (with its trees, buildings, and so on) acts as an obstruction to the wind. Hence, the wind slows over the town and deposits a portion of the blowing snow it was carrying. This of course, adds to the town's total snow accumulation.

4.   In clear weather there is usually a greater variation in atmospheric stability between morning and afternoon. Mornings are often characterized by a stable radiation inversion, and afternoons by an unstable atmosphere and strong vertical mixing. The greater atmospheric instability on clear days tends to link surface air with the air above and produce greater surface wind speeds. On overcast days the earth's surface does not warm as much, the atmosphere is more stable, and much less vertical mixing occurs.

5.   Since a city contains more concrete buildings and paved roads than the suburbs, the city would heat up during the day and be warmer than the surrounding suburbs. This would produce a daytime "suburb breeze" directed from the suburbs toward the city. At night, when the city cools, a "city breeze" would result.

6.   City A is the best site for a wind turbine. Here the winds appear to be steadier, yet strong enough to produce electricity throughout a 24-hour day.

7.   Site B is the best site to construct a wind turbine; Site C is the worst. The wind blows strongest at Site B and is most turbulent and gusty at Site C.

8.   Well-developed sea breezes are due to large daytime temperature and pressure differences that exist between the cool water and the warm land. If the water is quite cool, at night the land will be unable to cool to a temperature lower than that of the water. Consequently, at night pressure differences will be small and a land breeze will be unlikely.

9.   With a sea breeze, the air is rising over land and sinking over water. With a land breeze, the air is rising over water and sinking over land.

10.   The air moving inland over Florida is more unstable and contains more water vapor than does the air moving inland over California, off the cool Pacific Ocean. Only slight lifting will produce clouds in the humid, unstable Florida air.

11.   The fires are cooking breakfast. The sun is rising and heating the east-facing side of the hill. The air is rising on the warmer east-facing side and sinking on the cooler (shaded) west-facing side.

12.   The Appalachian Mountains are much lower in elevation than are the Rocky Mountains. Less dramatic warming of air occurs on the eastern side of the Appalachians. Also, the Appalachians do not effectively "shield" the eastern side from clouds, precipitation and humid air.

13.   For hot summer breezes from the north, the area should have a high topographic barrier, such as a high plateau, to the north. In addition, high surface pressure should be to the north. As the wind blows clockwise around the high, it moves southward (which is downhill). The sinking air is compressed, so it warms, producing high air temperatures.

14.   The strongest sea breezes should be along Florida's east coast. The strongest land breeze should be along Florida's west coast.

Answers to Problems and Exercises

2.   The total force exerted by the wind is 768 pounds.

Multiple Choice Exam Questions

1.   The smallest scale of atmospheric motion is the:
      a. mesoscale
      b. synoptic scale
      c. microscale
      d. macroscale
      e. global scale

ANSWER: c

2.   An example of mesoscale motion is:
      a. winds on a surface weather map of North America
      b. winds on a 500 mb chart
      c. winds blowing through a city
      d. winds blowing past a chimney
      e. average wind patterns around the world

ANSWER: c

3.   An example of microscale motion is:
      a. winds on a surface weather map of North America
      b. winds on a 500 mb chart
      c. winds blowing through a city
      d. winds blowing past a chimney
      e. average wind patterns around the world

ANSWER: d

4.   Which of the following associations is most accurate?
      a. microscale - chinook wind
      b. synoptic scale - sea breeze
      c. mesoscale - land breeze
      d. planetary scale - lake breeze

ANSWER: c

5. The slowing of the wind due to the <u>random motion</u> of air molecules is called:
    a. eddy viscosity
    b. mechanical turbulence
    c. molecular viscosity
    d. convective turbulence

ANSWER: c

6. Thermal turbulence above the surface is usually most severe:
    a. immediately after sunset
    b. at the time of maximum surface heating
    c. around midnight
    d. just before sunrise
    e. about midmorning, or soon after the minimum temperature is reached

ANSWER: b

7. On a clear, windy day, the depth to which mixing occurs above the surface depends upon:
    a. the wind speed
    b. surface heating
    c. the landscape
    d. all of the above

ANSWER: d

8. The top of the friction layer is usually found near what altitude?
    a. 100 m   (330 ft)
    b. 500 m (1,640 ft)
    c. 1000 m (3,300 ft)
    d. 5000 m (16,400 ft)

ANSWER: c

9. The friction of fluid flow is called:
    a. viscosity
    b. compression
    c. convergence
    d. shear
    e. inertia

ANSWER: a

10. Surface winds are generally strongest and most gusty:
    a. in the afternoon
    b. in the early morning
    c. around midnight
    d. just after sunset
    e. just before sunrise

ANSWER: a

11. The wind's speed generally increases with height above the earth's surface because:
    a. only the lowest layer of air rotates with the earth
    b. air temperature normally decreases with height
    c. wind instruments are not accurate at the earth's surface
    d. friction with the earth's surface slows the air near the ground
    e. air parcels expand and become less dense as they rise above the surface

ANSWER: d

12. The howling of wind on a blustery night is believed to be caused by:
    a. variation in air temperature
    b. wind blowing around a frozen surface
    c. snowflakes striking one another but not sticking together
    d. eddies of higher air density
    e. wind lifting, then lowering small objects at the surface

ANSWER: d

13. An abrupt change in wind speed or wind direction is called:
    a. wind shear
    b. an air pocket
    c. flurry
    d. squall

ANSWER: a

14. Violent, rotating eddies that create hazardous flying conditions beneath the crest of a mountain wave are called:
    a. mountainadoes
    b. dust devils
    c. rollers
    d. seiches
    e. rotors

ANSWER: e

15. Clear Air Turbulence (CAT) can occur:
    a. near a jet stream
    b. in areas of mountain waves
    c. where strong wind shear exists
    d. all of the above

ANSWER: d

16. Which below is <u>not</u> true concerning an "air pocket"?
   a. can form in a downdraft of an eddy
   b. often form in the atmosphere where the air is too thin to support the wings of an airplane
   c. often form in regions that exhibit strong vertical wind shear
   d. can develop in clear air

ANSWER: b

17. Pedaling a bicycle into a 20 knot wind will require about ___ as much effort as pedaling into a 10 knot wind.
   a. 2 times
   b. 4 times
   c. 10 times
   d. the same effort

ANSWER: b

18. The greatest wind speed ever recorded at the earth's surface occurred at:
   a. Mt. Washington, New Hampshire
   b. Miami, Florida
   c. Mt. Pleasant, Iowa
   d. Cheyenne, Wyoming
   e. Long's Peak, Colorado

ANSWER: a

19. Clear air turbulence often occurs near a boundary of high wind shear.
   a. true
   b. false

ANSWER: a

20. Clear air turbulence is typically a gentle phenomenon which poses little risk to aircraft.
   a. true
   b. false

ANSWER: b

21. Suppose the wind speed increased from 5 mph to 10 mph. We can conclude that the force exerted by the wind increased by a factor of
   a. 2
   b. 4
   c. 5
   d. 25

ANSWER: b

22. If huge waves pound against the beach communities of Southern California for several days during clear, calm weather, it is a good bet that:
a. the winds are strong somewhere out over the Pacific Ocean
b. an earthquake has occurred somewhere on the ocean floor
c. it is raining offshore
d. it is a period of high tides
e. there is a large difference in water density between the shoreline of Southern California and the middle of the Pacific

ANSWER: a

23. Which below has the <u>least</u> influence on determining how high a wind wave will grow over the open ocean?
a. fetch of water
b. density of water
c. length of time the wind blows over the water
d. wind speed over the water

ANSWER: b

24. Suppose the wind speed increased from 5 mph to 25 mph. We can conclude that the force exerted by the wind increased by a factor of
a. 2
b. 4
c. 5
d. 25

ANSWER: d

25. Which of the features below could indicate prevailing wind direction?
a. sand ripples
b. sand dunes
c. "flag" trees
d. all of the above

ANSWER: d

26. Dust storms and dust devils are _____ on Mars.
a. infrequent
b. extremely rare
c. common

ANSWER: c

27. A wind rose indicates:
a. the wind speed at a location at a particular time
b. the percentage of time that the wind blows from different directions
c. observed wind speed and direction on a surface map
d. spinning wind patterns caused by buildings or other obstructions

ANSWER: b

28. The most practical location for building a wind turbine would be:
    a. in a region of strong, gusty winds
    b. on the downwind side of a mountain
    c. in a narrow valley
    d. in a region of moderate, steady winds

ANSWER: d

29. An offshore wind:
    a. blows from land to water
    b. blows from water to land
    c. blows only at night
    d. only blows during the day

ANSWER: a

30. Which below would <u>not</u> be considered an onshore wind?
    a. sea breeze
    b. Santa Ana wind
    c. lake breeze
    d. none of the above

ANSWER: b

31. A wind reported as 045° would be a wind blowing from the:
    a. NE
    b. S
    c. SW
    d. NW
    e. E

ANSWER: a

32. Suppose a west wind of 20 knots blows over a coastal region which is densely covered in shrubs. If this same wind moves out over the middle of a large calm lake, its speed and direction would probably be:
    a. greater than 20 knots and more northwesterly
    b. less than 20 knots and more northwesterly
    c. greater than 20 knots and more southwesterly
    d. less than 20 knots and more southwesterly
    e. less than 20 knots and westerly

ANSWER: a

33. What instrument would you use for a rawinsonde observation?
    a. cup anemometer
    b. radiosonde
    c. aerovane
    d. wind profiler

ANSWER: b

34. An instrument used to measure wind speed is called a(an):
    a. anemometer
    b. ceilometer
    c. psychrometer
    d. tachometer

ANSWER: a

35. A wind reported as 225° would be a wind blowing from the:
    a. NE
    b. NW
    c. SE
    d. SW

ANSWER: d

36. A wind profiler obtains wind information using:
    a. an aerovane
    b. a theodolite
    c. a radiosonde
    d. a Doppler radar

ANSWER: d

37. A wind instrument that usually consists of 3 or more cups:
    a. wind vane
    b. aerovane
    c. wind profiler
    d. wind sock
    e. anemometer

ANSWER: e

38. The instrument that uses infrared or visible light in the form of a laser beam to determine wind direction is the:
    a. aerovane
    b. pilot balloon
    c. lidar
    d. radiosonde
    e. wind profiler

ANSWER: c

39. Which of the instruments below indicates both wind speed and wind direction?
    a. wind vane
    b. aerovane
    c. cup anemometer
    d. psychrometer
    e. theodolite

ANSWER: b

40.   The fact that the Martian surface _____ contributes to the high frequency of dust storms.
        a. is a global ocean
        b. is a global desert
        c. has no craters

ANSWER: b

41.   Which instrument only measures wind speed?
        a. anemometer
        b. wind sock
        c. wind vane
        d. rawinsonde observation
        e. aerovane

ANSWER: a

42.   Which is not a characteristic of a thermal low?
        a. forms in a region of warm air
        b. forms in response to variations in surface air temperature
        c. becomes stronger with increasing height
        d. lowest pressure is at the center

ANSWER: c

43.   If the sea level pressure in Philadelphia, Pennsylvania is 1016 mb, the strongest summertime sea breeze along the New Jersey coast would occur when the sea level pressure just east of Atlantic City, New Jersey is:
        a. 1019 mb
        b. 1017 mb
        c. 1016 mb
        d. 1015 mb

ANSWER: a

44.   During the summer in humid climates, nighttime clouds tend to form over water during a:
        a. land breeze
        b. chinook wind
        c. sea breeze
        d. Santa Ana wind

ANSWER: a

45.   Which below is usually not true concerning a sea breeze circulation?
        a. they mainly occur at night
        b. they usually occur when the water is cooler than the land
        c. they occur when the surface wind blows from the water toward the land
        d. can cause clouds to form over the land

ANSWER: a

46. A sea breeze circulation will reverse direction and become a land breeze:
    a. once every few days
    b. at the beginning and the end of the summer
    c. several times per day
    d. once per day

ANSWER: d

47. A sea or land breeze is caused by:
    a. differences in humidity
    b. temperature differences
    c. the Coriolis force
    d. ocean tides
    e. strong surf conditions

ANSWER: b

48. The nighttime counterpart of the sea breeze circulation is called a:
    a. chinook
    b. Santa Ana
    c. land breeze
    d. night breeze
    e. foehn

ANSWER: c

49. In summer, during the passage of a sea breeze, which of the following is not usually observed?
    a. a drop in temperature
    b. a drop in relative humidity
    c. a wind shift
    d. an increase in relative humidity

ANSWER: b

50. A smog front is most often associated with which wind system?
    a. monsoon
    b. chinook
    c. Santa Ana
    d. mountain breeze
    e. sea breeze

ANSWER: e

51. A cool, summertime wind that blows from sea to land is called a:
    a. Santa Ana wind
    b. land breeze
    c. valley breeze
    d. sea breeze

ANSWER: d

52. In south Florida the prevailing winds are northeasterly. Because of this, the strongest sea breeze is usually observed on Florida's _____ coast, and the strongest land breeze on Florida's ___ coast.
    a. east, west
    b. west, east
    c. west, west
    d. east, east

ANSWER: a

53. When a sea breeze moving north meets a sea breeze moving south they form a:
    a. land breeze
    b. sea breeze convergence zone
    c. katabatic wind
    d. valley breeze

ANSWER: b

54. Clouds and precipitation are frequently found on the downwind side of a large lake. This would indicate that the air on the downwind side is:
    a. converging and rising
    b. converging and sinking
    c. diverging and sinking
    d. diverging and rising

ANSWER: a

55. During the summer along the coast, a sea breeze is usually strongest and best developed:
    a. in the afternoon
    b. just after sunrise
    c. just before sunset
    d. just before noon
    e. around midnight

ANSWER: a

56. Monsoon depressions are:
    a. huge drainage gullies that are produced during the heavy rains in the summer monsoon
    b. upper-level jet streams
    c. low pressure areas that enhance rainfall during the summer monsoon
    d. large reservoirs used for irrigation that fill with water during the summer monsoon
    e. a period of generally good weather with lower-than-average rainfall that may last for several days during the otherwise rainy summer monsoon

ANSWER: c

57.   The summer monsoon in eastern and southern Asia is characterized by:
        a. wet weather and winds blowing from land to sea
        b. dry weather and winds blowing from land to sea
        c. wet weather and winds blowing from sea to land
        d. dry weather and winds blowing from sea to land

ANSWER: c

58.   An extremely strong downslope wind that occurs in parts of Antarctica would be considered a
        a. katabatic wind
        b. Santa Ana wind
        c. mountain breeze
        d. monsoon circulation
        e. diurnal breeze

ANSWER: a

59.   The winter monsoon in eastern and southern Asia is characterized by:
        a. wet weather and winds blowing from land to sea
        b. wet weather and winds blowing from sea to land
        c. dry weather and winds blowing from sea to land
        d. dry weather and winds blowing from land to sea

ANSWER: d

60.   Low pressure becomes best developed over the Asian continent in:
        a. summer
        b. winter
        c. fall
        d. spring

ANSWER: a

61.   While fly fishing in a mountain stream, you notice that the wind is blowing upstream. From this you could deduce that the wind is a:
        a. chinook wind
        b. valley breeze
        c. Santa Ana wind
        d. mountain breeze
        e. katabatic wind

ANSWER: b

62.   Cumulus clouds that appear above isolated mountain peaks are often the result of:
   a. katabatic winds
   b. mountain winds
   c. fall winds
   d. Santa Ana winds
   e. valley breezes

ANSWER: e

63.   A valley breeze would develop its maximum strength:
   a. at sunrise
   b. in early afternoon
   c. about an hour after sunset
   d. about midnight

ANSWER: b

64.   Which of the winds below is not considered to be a cold wind?
   a. Texas norther
   b. mistral
   c. California norther
   d. bora
   e. blizzard

ANSWER: c

65.   A strong, usually cold, downslope wind is called a:
   a. valley wind
   b. katabatic wind
   c. monsoon wind
   d. haboob wind
   e. chinook wind

ANSWER: b

66.   Which of the following is not considered to be a katabatic wind?
   a. bora
   b. mistral
   c. haboob
   d. mountain breeze
   e. fall wind

ANSWER: c

67.   A katabatic wind on the Oregon coast would most likely blow from the:
   a. north
   b. south
   c. east
   d. west

ANSWER: c

68. A katabatic wind is a ___, ___ wind.
    a. cold, upslope
    b. warm, upslope
    c. cold, downslope
    d. warm, downslope

ANSWER: c

69. A chinook wind in the Alps is called a:
    a. haboob
    b. monsoon
    c. foehn
    d. bora
    e. Santa Ana

ANSWER: c

70. The heat from a chinook wind is generated mainly by:
    a. compressional heating
    b. sunlight
    c. warm, ocean water
    d. friction with the ground
    e. forest fires

ANSWER: a

71. A chinook wall cloud is a:
    a. sand storm that marks the leading edge of the chinook
    b. row of intense thunderstorms that bring heavy rain to eastern Colorado
    c. cloud of smoke the marks the advancing edge of the chinook
    d. bank of clouds that form over the mountains and signal the possible onset of a chinook
    e. line of fog that moves over the plains as a chinook advances

ANSWER: d

72. On the eastern side of the Rocky Mountains, chinook winds are driest when:
    a. clouds form and precipitation falls on the upwind side of the mountains
    b. the air aloft is cold
    c. the sun is shining
    d. the winds are blowing from the east
    e. surface friction is greatest on the downwind side of the mountain

ANSWER: a

73. Chinook winds are:
    a. warm, dry downslope winds
    b. warm, moist downslope winds
    c. cold, dry downslope winds
    d. cold, moist downslope winds
    e. warm, dry upslope winds

ANSWER: a

74. The main reason Santa Ana winds are warm is because:
    a. latent heat is released in rising air
    b. sinking air warms by compression
    c. condensation occurs
    d. solar heating warms the air
    e. they are heated by forest fires in canyons

ANSWER: b

75. The Santa Ana wind is a ___, ___ wind that blows into southern California.
    a. cold, damp
    b. cold, dry
    c. warm, moist
    d. warm, dry

ANSWER: d

76. Strong Santa Ana winds develop in Los Angeles during the fall when a ___ pressure center forms to the ___ .
    a. high, northeast of Los Angeles over the Great Basin
    b. high, southwest of Los Angeles over the Pacific Ocean
    c. low, northeast of Los Angeles over the Great Basin
    d. low, southwest of Los Angeles over the Pacific Ocean

ANSWER: a

77. Which of the following would not be considered a "desert" wind?
    a. norte
    b. leste
    c. khamsin
    d. sharav
    e. leveche

ANSWER: a

78. A warm, dry gusty wind that blows across north Africa is the:
    a. burga
    b. buran
    c. levantar
    d. tehauntepecer
    e. sirocco

ANSWER: e

79. Which of the following conditions favor the development of dust devils?
    a. hot, moist days
    b. hot, dry days
    c. cold, moist days
    d. cold, dry days

ANSWER: b

80. Another name for a small, rotating whirlwind observed at the surface is:
    a. seiche
    b. haboob
    c. rotor
    d. dust devil
    e. foehn

ANSWER: d

81. A dust or sandstorm that forms along the leading edge of a thunderstorm is a:
    a. foehn
    b. haboob
    c. chinook
    d. bora
    e. Santa Ana

ANSWER: b

82. A northeaster along the east coast of the United States is best developed when a low pressure area:
    a. moves from west to east over Canada
    b. moves eastward over the Great Lakes
    c. moves northward over Ohio
    d. moves northeastward from the Texas panhandle
    e. moves northeastward over the Atlantic Ocean adjacent to the coast

ANSWER: e

83. A sea breeze is most likely to develop along a coastline when
     a. away from the shore, pressure gradients are large.
     b. away from the shore, pressure gradients are small.
     c. it is raining.
     d. it is foggy.

ANSWER: b

Essay Exam Questions

1. You are hiking on a mountain trail at sunrise when you smell the smoke from cooking bacon. You can't see where the smoke is coming from. Would you expect the camp to be above you or below you on the mountain? Explain.

2. List the scales of atmospheric motion from largest to smallest and give an example of each. From what you know about the various types of wind systems, are the size and duration related?

3. What is clear air turbulence (CAT)? Why does clear air turbulence represent a hazard to aviation?

4. Draw a sketch to show where eddies can form when air blows rapidly over a mountain range. Show on your sketch where you might expect clouds to form. How would these clouds appear when viewed from the ground?

5. Summertime weather forecasts for cities on the shores of the Great Lakes often contain the phrase "cooler near the lake". Explain the relevance of this phrase.

6. How might a knowledge of the direction of the prevailing wind at a given location be used in landscape design or the design of an energy efficient home?

7. Would you expect a well-developed sea breeze circulation to cause clouds to form over the land or over the ocean?

8. Briefly sketch or describe the formation of a chinook wind. Would you expect chinook winds to form more often on the eastern or the western slopes of the Cascade mountains in Oregon?

9. Briefly sketch or describe the conditions that lead to the formation of a Santa Ana wind. Is the Santa Ana wind dry or moist, warm or cold?

10. What is meant by the term monsoon wind system? Briefly describe or sketch the wind and pressure pattern during the summer and winter monsoon in Asia.

11. Will a valley breeze or a mountain breeze produce clouds? Explain.

12. Explain how Florida's sea breezes can increase the potential for wildfires.

Chapter 11
Wind: Global Systems

Summary

The study of winds continues in this chapter with a discussion of global scale circulation pattern

The three-cell model of the general circulation is presented and is compared with real world observations of pressures and winds. The seasonal movement of these features and their effect on regional climate is studied.

The chapter examines the major jet streams found in the earth's atmosphere. The polar jet stream and the subtropical jet form in regions with strong horizontal temperature gradients. Conservation of angular momentum also plays a role in the formation of these fast moving currents of air. Interactions between global scale wind patterns and the earth's oceans are examined, and the positions and directions of the major ocean currents are shown. Finally, the El Niño/Southern Oscillation, the North Atlantic Oscillation and the Pacific Decadal Oscillation are described, with considerable attention given to the possible climatological effects associated with both El Niño and La Niña events.

Key Terms

general circulation (of
   the atmosphere)
single-cell model
Hadley cell
three-cell model
doldrums
convective "hot" towers
subtropical highs
horse latitudes
trade winds
northeast trades
southeast trades
intertropical
   convergence zone
   (ITCZ)
westerlies
polar front
subpolar low
Ferrel cell
polar easterlies
General Circulation
   Models (GCMs)
semipermanent highs
   and lows
Bermuda high

Pacific high
Icelandic low
Aleutian low
Siberian high
thermal lows
monsoon low
"dishpan" experiment
jet streams
subtropical jet stream
polar front jet stream
tropopause jets
isotachs
jet maximum
   (or jet streak)
subtropical front
conservation of
   angular momentum
linear momentum
angular momentum
tropical easterly
   jet stream
stratospheric polar night
   jet stream
low-level jet
gyre

Gulf Stream
Labrador Current
North Atlantic Current
Canary Current
North Equatorial
   Current
Kuroshio Current
North Pacific Drift
California Current
oceanic front
Ekman Spiral
upwelling
El Niño
major El Niño event
Kelvin wave
Southern Oscillation
El Niño/Southern
   Oscillation
ENSO ·
Pineapple connection
La Niña
TOGA (Tropical Ocean
   and Global Atmosphere)
WCRP (World Climate
   Research Program)

Teaching Suggestions

1.   Clouds produced by some of the mesoscale and global scale wind circulation patterns discussed in this chapter can be seen on satellite photographs. The ITCZ is often very clearly defined on a full or half disk photograph from a geostationary satellite.

2.   During the winter, examine a surface weather map in class and compare temperatures to the north and the south of the polar front. There is often a very sharp temperature gradient. Locate the polar jet stream on an upper-level chart and determine the region where the maximum winds are found.

Student Projects

1.   Have students research and plot the path of some of the early voyages of discovery made aboard sailing ships (Columbus or Magellan, for example). Do the routes appear reasonable in light of what the students know about the earth's general circulation pattern?

2.   Have students collect climatological data for their location (or another region) during one or two strong El Niño events. Select two or three periods when there was not a strong El Niño to act as a control.

What effects might the El Niño have on local weather or climate? Do the students see any evidence of this in their climate data?

Blue Skies 3.   Use the Atmospheric Forces/Winds in Two Hemispheres section of the BlueSkies cdrom to try to locate some of the major features of the global wind circulation. Which features can you find? In which part(s) of the world did you find them?

Blue Skies 4.   Use the Weather Forecasting/Forecasting section of the BlueSkies cdrom to look for jet streams on an upper-level weather map. Name any jet streams that you find.

Answers to Questions For Thought

1.   Continents would disrupt the flow by creating several smaller circulations.

2.   The position of the major features of the general circulation would change. In summer, the subtropical highs and ITCZ would be displaced farther north than now; in the winter, the subtropical highs and polar front would move farther south than they now do.

3.   The surface low is a shallow, thermal low due to hot surface air; it disappears at upper levels. The subtropical high shows up clearly on upper-level charts such as those at 850 mb, 700 mb, and 500 mb.

4.   Because of the Ekman Spiral, the average movements of surface water down to a depth of about 100 m is at right angles to the surface wind direction. Icebergs, which may extend downward to depths greater than 100 m, move with this surface water at nearly right angles to the surface wind direction.

5.   The subsiding air of the subtropical Pacific high carries ozone downward from the stratosphere.

6.   The fastest winds are found in the jet stream core. Clear air turbulence is found above and below the jet stream core.

7.   This is due mainly to the reversal of winds associated with the summer and winter monsoon.

8.   Upwelling is strongest in summer and when the winds blow parallel to the coast. Strong upwelling conditions bring cold water to the surface. In winter when upwelling is not as strong, the surface water is not as cold.

9.   The subtropical jet stream forms to the south of the polar-front jet.

10.   Large subtropical highs tend to be centered over the oceans off the western margin of continents in both hemispheres. Winds around the highs blow clockwise (from the north along the western margin of the continents) in the Northern Hemisphere, and counterclockwise (from the south along the western margin of continents) in the Southern Hemisphere.

11.   Lake Michigan is located in the midlatitudes, and therefore is in an area of prevailing westerly winds. The westerlies push the warm surface water towards the eastern shore of the lake, creating comfortable swimming conditions. Along the western shore, upwelling brings cold water to the surface,

creating chilly swimming conditions. [Note: in the text, the word "cooler" should be replaced with "warmer".]

Answers to Problems and Exercises

1.

	July	January
(a)	S	NW to NNW
(b)	SSW	NW to NNW
(c)	WNW	SE
(d)	WNW to NW	SW
(e)	WNW to NW	SW
(f)	SW	S
(g)	W to WSW	S to SSE
(h)	N to NNW	W to WSW

2.   Time = distance/velocity. With an average velocity of 100 kts (=51 m/s) and a distance of 3000 km (=3 x 10^6 m), the travel time would be 58,824 seconds or 16.3 hours.

3.   The subtropical Pacific high pressure area east of California causes the prevailing winds to be northwesterly in San Francisco. The winds, in turn, cause upwelling along the coast. The high pressure area also keeps the air aloft stable, so that advection fog and low clouds are able to form near the surface and, at the same time, precipitation is rare. The cool water produces low maximum and minimum daily temperatures in San Francisco. The dew point is near that of the water temperature.

In Atlantic City, New Jersey, the winds tend to be southerly as air flows northward around the Bermuda high. The air from the south is both warm and humid. Temperatures and dew points are high, the air is unstable, and the water temperature is fairly high as the Gulf Stream moves in close to shore. Thunderstorms are possible in the humid unstable air of the afternoon and, hence, precipitation is common.

Multiple Choice Exam Questions

1.   A westerly wind means that the atmosphere:
a. would act to decrease the rotation of the earth
b. is moving faster than the earth spins
c. is moving slower than the earth spins
d. is moving as fast as the earth rotates

ANSWER: b

2.	Which below is <u>not</u> an assumption of the single-cell model of the general circulation of the atmosphere?
	a. the earth's surface is covered with water
	b. the earth rotates once in 24 hours
	c. the sun is always overhead at the equator
	d. none of the above

ANSWER: b

3.	The large thermally driven convection cell that is driven by convective "hot" towers along the equator is the:
	a. Ferrel cell
	b. Hadley cell
	c. Ekman spiral
	d. El Niño cell

ANSWER: b

4.	Air moving eastward more slowly than the earth rotates, would appear to be ____ wind to an observer on the earth.
	a. a counterclockwise
	b. an east
	c. an upward
	d. a west

ANSWER: b

5.	Chicago, Illinois (latitude 42° N) is located in the ___.
	a. northeast trades
	b. southeast trades
	c. westerlies
	d. doldrums

ANSWER: c

6.	The intertropical convergence zone (ITCZ) is a region where:
	a. the polar front meets the subtropical high
	b. northeast trades meet the southeast trades
	c. northeast trades converge with the subtropical high
	d. the Ferrel cell converges with the Hadley cell
	e. polar easterlies converge with the air at the doldrums

ANSWER: b

7.	In terms of the three-cell model of the general circulation, areas of surface low pressure should be found at:
	a. the equator and the poles
	b. the equator and 30° latitude
	c. the equator and 60° latitude
	d. 30° latitude and 60° latitude
	e. 30° latitude and the poles

ANSWER: c

8.	In Honolulu, Hawaii (latitude 21° N), you would most likely experience winds blowing from the:
	a. northeast
	b. south
	c. southwest
	d. northwest

ANSWER: a

9.	At Barrow, Alaska (latitude 70° N), you would expect the prevailing wind to be:
	a. northerly
	b. easterly
	c. southerly
	d. westerly

ANSWER: b

10.	Generally, along the polar front one would not expect to observe:
	a. temperatures on one side lower than on the other side
	b. an elongated region of lower pressure
	c. clouds and precipitation
	d. converging surface air
	e. sinking air aloft

ANSWER: e

11.	According to the three-cell general circulation model, at the equator we would not expect to find:
	a. the ITCZ
	b. a ridge of high pressure
	c. cumuliform clouds
	d. light winds
	e. heavy showers

ANSWER: b

12. The wind belt observed on the poleward side of the polar front is called the:
     a. polar easterlies
     b. prevailing westerlies
     c. northeast trades
     d. doldrums

ANSWER: a

13. On a weather map of the Northern Hemisphere, one would observe the westerlies:
     a. north of the subpolar lows
     b. south of the tropical highs
     c. between the doldrums and the horse latitudes
     d. between the subpolar lows and the subtropical highs

ANSWER: d

14. The majority of the United States lies within this wind belt:
     a. westerlies
     b. easterlies
     c. northerlies
     d. trades
     e. southerlies

ANSWER: a

15. On a weather map of the Northern Hemisphere, the trade winds would be observed:
     a. north of the polar front
     b. between the polar front and the subtropical highs
     c. south of the subtropical highs
     d. between the subpolar lows and the subtropical highs

ANSWER: c

16. In the general circulation of the atmosphere, one would find the region called the doldrums:
     a. near 30° latitude
     b. at the equator
     c. at the poles
     d. near 60° latitude

ANSWER: b

17. In terms of the three-cell general circulation model, the driest regions of the earth should be near:
    a. the equator and the polar regions
    b. the equator and 30° latitude
    c. the equator and 60° latitude
    d. 30° latitude and 60° latitude
    e. 30° latitude and the polar regions

ANSWER: e

18. The world's deserts are found at 30° latitude because:
    a. the intertropical convergence zone is located there
    b. of the sinking air of the polar front
    c. of the convergence of the prevailing westerlies and the Northeast Trades
    d. of the sinking air of the subtropical highs
    e. of the doldrums

ANSWER: d

19. On a <u>surface weather map</u> during the month of July, one would expect to find what type of pressure system over the desert southwest of the United States?
    a. monsoon low
    b. Pacific high
    c. thermal low
    d. Aleutian low

ANSWER: c

20. The semi-permanent pressure systems associated with the polar front are called:
    a. subpolar lows
    b. equatorial lows
    c. polar highs
    d. subtropical highs

ANSWER: a

21. Which of the following is not considered a semi-permanent high or low **pressure area**?
    a. Bermuda high
    b. Aleutian low
    c. Siberian high
    d. Pacific high
    e. Icelandic high

ANSWER: c

22. The position of the Pacific high over the north Pacific Ocean shifts ___ in winter and ___ in summer.
    a. northward, southward
    b. southward, northward
    c. eastward, westward
    d. westward, eastward

ANSWER: b

23. The large semi-permanent surface anticyclone that is normally positioned over the ocean, west of California, is called the:
    a. Hawaiian high
    b. Aleutian high
    c. California high
    d. Baja high
    e. Pacific high

ANSWER: e

24. A thermally direct cell is one that
    a. is found in both the northern and southern hemispheres
    b. involves heat
    c. occurs on both sides of a large lake or ocean
    d. is driven by energy from the sun

ANSWER: d

25. Many small-scale processes are _____ in General Circulation Models.
    a. ignored
    b. approximated
    c. parameterized
    d. both b and c
    e. none of the above

ANSWER: d

26. Which of the following does not describe the subtropical jet stream?
    a. forms along the polar front
    b. generally blows from west to east
    c. is found at the tropopause
    d. is normally equatorward of the polar front jet stream

ANSWER: a

27.   Which below does not describe the polar front jet stream?
        a. is strongest in winter
        b. moves farther south in winter
        c. forms near the boundary called the polar front
        d. is normally found at a higher elevation than the subtropical jet

ANSWER: d

28.   In the Northern Hemisphere, the polar jet stream is strongest when:
        a. air north of the polar front is much colder than air south of the polar front
        b. air north of the polar front is much warmer than air south of the polar front
        c. air temperatures on opposite sides of the polar front are about equal
        d. air temperatures on the East Coast of the US are much colder than on the West
           Coast of the US

ANSWER: a

29.   The jet stream flows:
        a. directly from west to east
        b. directly from east to west
        c. from the equator towards the poles
        d. in a wavy pattern from west to east

ANSWER: d

30.   As an air parcel aloft moves northward from the equator, it moves closer to the earth's axis of
        rotation. Because of the conservation of angular momentum, the air parcel's motion should:
        a. remain constant
        b. slow and eventually reverse direction
        c. slow but continue in the same direction
        d. increase in speed

ANSWER: d

31.   In the Northern Hemisphere, air found to the north of the polar front
        is ___ , while air further south is ___ :
        a. cold, warm
        b. cold, cold
        c. warm, cold
        d. warm, warm

ANSWER: a

32.   The average winds aloft are strongest in:
       a. summer
       b. winter
       c. fall
       d. spring

ANSWER: b

33.   The low-level jet that forms over the Central Plains of the United States appears to be responsible for:
       a. the influx of cold, polar air
       b. the development of El Niño
       c. the onset of a chinook
       d. the formation of a subtropical high pressure area
       e. nighttime thunderstorms

ANSWER: e

34.   Factors that contribute to the formation of a low-level jet stream over the Central Plains of the United States are:
       a. the sloping of the land from the Rockies to the Mississippi Valley
       b. a north-south trending mountain range
       c. stable air above the jet
       d. all of the above

ANSWER: d

35.   Which below is not correct about the stratospheric polar-night jet stream?
       a. it is strongest in winter
       b. it blows from west to east
       c. it is found near the tropopause
       d. it is caused by a north-south temperature gradient

ANSWER: c

36.   Generally, on an upper-level (100 mb) chart in the Northern Hemisphere during July we would expect to find the tropical easterly jet stream:
       a. north of the polar front
       b. north of the upper-level subtropical high
       c. directly over the polar front
       d. south of the upper-level subtropical high
       e. in the upper stratosphere

ANSWER: d

37.   A phenomenon in the Atlantic Ocean, similar to the southern oscillation, is the
        a. Florida sea breeze convergence front
        b. Asian monsoon
        c. North Atlantic Oscillation

ANSWER: c

38.   Average winter temperatures in Great Britain and Norway would probably be much colder if it
        were not for the:
        a. Labrador current
        b. North Atlantic Drift
        c. Canary current
        d. North Equatorial current
        e. Greenland current

ANSWER: b

39.   In the Northern Hemisphere, ocean currents in the Atlantic and the Pacific move in a generally
        circular pattern. The direction of this motion is __ in the Atlantic and __ in the Pacific.
        a. clockwise, counterclockwise
        b. counterclockwise, counterclockwise
        c. clockwise, clockwise
        d. counterclockwise, clockwise

ANSWER: c

40.   The Pacific Decadal Oscillation is similar to the El Niño/Southern Oscillation, except that it
        a. also involves ocean currents
        b. affects a large area of the earth
        c. reverses every 20 to 30 years
        d. affects fish populations

ANSWER: d

41.   At any given time, only one jet stream can be found in the atmosphere.
        a. true
        b. false

ANSWER: b

42.   The cold water observed along the northern California coast in summer is due mainly to:
        a. the California current
        b. oceanic fronts
        c. upwelling
        d. cold air moving over the water
        e. evaporation

ANSWER: c

43.   Major ocean currents that flow parallel to the coast of North America are:
        a. Labrador, Canary, California
        b. California, Gulf Stream, Labrador
        c. Kuroshio, California, Labrador
        d. Labrador, Canary, Gulf Stream

ANSWER: b

44.   The name given to the current of warm water that replaces cold surface water along the coast of Peru and Ecuador during December is:
        a. Brazil current
        b. Humbolt current
        c. Benguela current
        d. El Niño

ANSWER: d

45.   The two ocean currents, warm and cold, that produce fog off the coast of Newfoundland are the:
        a. Gulf stream and Canary current
        b. Labrador current and Canary current
        c. Gulf stream and Labrador current
        d. North Atlantic Drift and Canary current
        e. North Atlantic Drift and Gulf stream

ANSWER: c

46.   The turning of water with depth is known as:
        a. the Hadley Spiral
        b. the Ekman Spiral
        c. a gyre
        d. the Ferrel Spiral
        e. upwelling

ANSWER: b

47.   The Ekman Spiral describes:
        a. the turning of water with depth
        b. the air flow into a center of low pressure
        c. the circulation of surface water around a gyre
        d. the air flow out of a region of high pressure
        e. the wind-flow pattern in a jet stream

ANSWER: a

48.　Upwelling is:
　　a. the lifting of air along the polar front
　　b. the rising of cold water from below
　　c. increasing heights in an upper-level ridge
　　d. the rising air motion found in a low pressure center

ANSWER: b

49.　During a major El Niño event:
　　a. Peruvian fishermen harvest a record amount of fish near Christmas time
　　b. extensive ocean warming occurs over the tropical Pacific
　　c. the Northeast trade winds increase in strength
　　d. California experiences severe drought conditions

ANSWER: b

50.　At jet streak is a place where _____ is often found.
　　a. clear air turbulence
　　b. strong vertical wind speed shear
　　c. very strong winds
　　d. all of the above

ANSWER: d

51.　Upwelling occurs along the northern California coast because:
　　a. winds cause surface waters to move away from the coast
　　b. of seismic activity on the ocean bottom
　　c. of gravitational attraction between the earth and the moon
　　d. water flows from the Atlantic ocean into the Pacific because they are at
　　　　different levels

ANSWER: a

52.　A condition where the central and eastern tropical Pacific Ocean turns cooler than normal is
　　called:
　　a. El Niño
　　b. La Niña
　　c. the Southern Oscillation
　　d. the Ekman Spiral

ANSWER: b

53. The reversal of the positions of surface high and low pressure at opposite ends of the Pacific Ocean is called:
    a. El Niño
    b. the Southern Oscillation
    c. upwelling
    d. La Niña

ANSWER: b

54. If the earth's surface was homogeneous (either all land or all water),
    a. the semipermanent highs and lows would be stronger
    b. the semipermanent highs and lows would be weaker
    c. the semipermanent highs and lows would disappear altogether
    d. the semipermanent highs and lows wouldn't change in intensity

ANSWER: c

Essay Exam Questions

1. On a large circle, show where the major pressure and wind belts would be found according to the 3-cell model of the earth's general circulation.

2. What changes might you expect to see in the earth's general circulation if the earth's rotation were in the opposite direction?

3. Briefly describe where each of the following features is found in the earth's general circulation. What meteorological conditions might you find associated with each feature?
   horse latitudes   trade winds   ITCZ
   polar front   doldrums

4. If you had to build a bridge linking South America to Africa, at what latitude would you build it? Why?

5. What occurs during a major El Niño event? How might this affect global wind and precipitation patterns?

6. What is the polar jet stream and where is it found? Does the position of the polar jet stream change during the year?

7. What features in the earth's general circulation help determine where the driest and wettest places on earth are found?

8. Explain why large high and low pressure systems don't move from east-to-west across the continental United States.

Chapter 12
Air Masses and Fronts

Summary

This chapter examines the typical weather conditions associated with air masses and the weather produced at frontal boundaries between air masses.

Students will first see how and where air masses form and how they are classified according to their temperature and humidity properties. Once upper level winds cause an air mass to move, the air mass will carry characteristics of its source with it and may have a strong influence on conditions in the region it invades. Continental polar air moving down from Canada, for example, often brings clear skies but bitterly cold temperatures to the United States in winter.

The converging air motion around areas of low pressure will often bring air masses with widely different properties into contact. The different types of fronts that form at the boundaries between air masses are discussed next in the chapter. When warm and cold air masses move toward each other, the warm, low-density air is forced upward. The rising motion is most gradual in the case of a warm front, and precipitation can occur over a large area ahead of the front. Air is generally forced upward more abruptly at cold fronts with the result that precipitation may be quite heavy in a narrower zone near the front. Typical weather conditions that might be observed during the approach and passage of warm, cold, and occluded fronts are summarized.

The meteorological phenomena of drylines and upper-air fronts are also described in this chapter. Drylines, or dewpoint fronts, are sometimes associated with severe weather, while upper-air fronts are found in conjunction with tropopause folds and downward incursions of stratospheric air.

Key Terms

air mass
source regions (for
  air masses)
continental polar (cP)
continental arctic (cA)
Texas norther
lake-effect snows
fetch
polar outbreak
Siberian Express
maritime polar (mP)
Pacific air

"back door" cold front
 warm front
overrunning
maritime tropical (mT)
continental tropical (cT)
air mass weather
front
frontal surface
frontal zone
stationary front
cold front

pressure tendency
trough
squall line
frontolysis
frontogenesis
frontal inversion
veering (wind)
occluded front
 (occlusion)
cold occlusion
warm occlusion

Teaching Suggestions

1.   A satellite photograph of the spotty pattern of cumuliform clouds or rows of clouds that are produced when continental polar air moves out over warm ocean water can be used to complement a discussion of the lake effect. Ask the students what type of weather conditions they would expect if the winds in such a photograph changed direction and began to blow from the ocean toward the land.

2.   Show the students a good example of the "comma-shaped" cloud pattern associated with a mature middle latitude storm. Ask the students where they would expect the center of low pressure and the fronts to be found.

A line of thunderstorms forming along a strong cold front can often be seen clearly on satellite photographs.

3.   After covering the material in this chapter, students should be able to understand and enjoy discussions of the current weather conditions depicted on surface weather charts. Show the positions and movement of air masses and fronts. Show the upper air chart and relate this to the surface features.

4.   Students will sometimes be confused to find precipitation associated with a stationary front. It is worth explaining that warm air may still override the cold air even though the cold air mass and the frontal boundary remain stationary.

Student Projects

1.   Provide the students with surface weather observations plotted on a map. Have the students first locate centers of high and low pressure. Then using the weather changes summarized in Tables 12.2 and 12.3 have them attempt to locate warm and cold fronts on the map. The instructor can supply students with simple examples at first and then move to more complex situations. Having students forecast the future movement of a middle latitude storm would fit in well with material covered in Chapter 14.

Have students draw isotherms on the surface weather chart. A southward bulge of cold air will often be visible to the west of a strong surface low pressure center. The cold front should correspond to the front edge of this cold air mass. Similarly, the warm front will be found at the advancing edge of a warm air mass east of the low.

2.   Have students record and plot daily average weather data (maximum and minimum temperature, average dew point and pressure, precipitation amounts) and weather observations (cloud cover, cloud types, winds) for a few days before and following the passage of a strong front. The change in weather conditions can sometimes be quite dramatic. Also, students will often be surprised to see the sequence of events described in the text actually occurring in the real world.

Students could repeat this same exercise for another location.

3.   Have students describe and document an unusual weather event that occurs during the semester, such as an outbreak of polar air, a squall line with severe thunderstorms in the southeastern US, a strong storm with gale force winds reaching the northwestern US, or a strong storm along the East Coast of the US. The study should be confined to air mass weather or a middle latitude storm system. The student's report should include a surface weather map and an upper level map. In each case students should attempt to find one or more reasons for these extreme weather conditions. Was the central pressure in a surface low, for example, lower than normal? Was the temperature gradient across a cold front unusually large? Was the upper level wind flow pattern atypical?

Students might also document an unusual weather event that they or someone from their family remembers.

Blue Skies 4.   Use the Weather Analysis/Isoplething and Weather Analysis/Find the Front activities on the BlueSkies cdrom to demonstrate isoplething and frontal analysis. Discuss any fronts in connection with local weather changes. What variables are most important for identifying this front?

Blue Skies 5.   Use the Weather Forecasting/Forecasting activity on the BlueSkies cdrom to determine what type of air mass is affecting your area today.

Answers to Questions for Thought

1.   The mP air mass would dry considerably as it crosses the Rockies. In winter, on the eastern side of the Rockies, the air mass would probably be warmer than the air it replaces. The air mass would also become more stable as it moves over the cold ground. Precipitation would probably occur along the leading edge of the air mass, especially if Gulf air moves northward ahead of it. In summer, an mP air mass on the eastern side of the Rockies would probably be cooler than the air it replaces. Heating the surface would make it unstable in the lowest layers and cumulus clouds might form. Also, along the advancing edge of the air mass, showers and thunderstorms would form as warm, humid air is forced to rise.

2.   On the eastern side of the anticyclone the winds are northerly. Cold winds and cP air can bring record low temperatures. On the western side of the anticyclone the winds are southerly. As the anticyclone drifts eastward, the southerly winds carry warm mT air into the region.

3.   The temperature inversion in Fig. 12.4 appears to be a radiation (surface) inversion. The inversions in Fig. 12.18b are frontal inversions due to warm air overrunning cold surface air.

4.   North-northeast or northeast.

5.   When a very cold air mass moves out of Canada into the United States the flow aloft is meridional and the upper-level pattern is wavy. Hence, the northerly winds aloft direct a cold air mass into one part of the United States while, at the same time, the southerly winds aloft direct a warm air mass into another portion of the United States.

6.   Freezing rain is more common with warm fronts because with a warm front, warm air rides up and over cold surface air. It is the warm rain falling into the cold, stable surface air that produces freezing rain. Often, behind a cold front the air becomes colder aloft.

7.   In winter, cold fronts are well developed. When warm, humid mT air is drawn northward ahead of the front the warm air is lifted, often producing stormy weather. In winter, along a warm front the air is usually stable as warm air lies above cold air. In summer, along a warm front, warm, humid, unstable air rides up and over only slightly cooler surface air. Often the rising unstable air is able to produce towering clouds, showers, and even thunderstorms.

8.   Counterclockwise.

9.   As a cold front moves eastward, mT air is drawn up from the Gulf of Mexico ahead of it.

10.   A warm front.

11.   Very unlikely, since the lake wouldn't be expected to thaw in February.

Answers to Problems and Exercises

2.   The warm front would be about 1200 km (750 miles) away. It should pass the area in about 65 hours (1 knot = 1.15 mph).

Multiple Choice Exam Questions

1.   A good source region for an air mass would be:
        a. mountains with deep valleys and strong surface winds
        b. generally flat areas of uniform composition with light surface winds
        c. hilly with deep valleys and light winds
        d. generally flat area of uniform composition with strong surface winds

ANSWER: b

2.   The origin of cP and cA air masses that enter the United States is:
        a. Northern Siberia
        b. Northern Atlantic Ocean
        c. Antarctica
        d. Northern Canada and Alaska

ANSWER: d

3.   Which of the following statements is most plausible?
        a. In winter, cP source regions have higher temperatures than mT source regions
        b. In summer, mP source regions have higher temperatures than cT source regions
        c. In winter, cA source regions have lower temperatures than cP source regions
        d. In summer, mT source regions have lower temperatures than mP source regions
        e. They are all equally plausible

ANSWER: c

4.   The temperature that unsaturated air would have if it moved from its original altitude to a
pressure of 1000 mb is the
        a. dew point temperature
        b. wet bulb temperature
        c. potential temperature
        d. adiabatic temperature

ANSWER: c

5.   In an exceptionally cold winter during which the Great Lakes were entirely covered by ice, lake
        effect snows would be expected in extremely high frequency and intensity.
        a. true
        b. false

ANSWER: b

6.   Compared to an mP air mass, mT air is ___.
        a. warmer and drier
        b. warmer and moister
        c. colder and drier
        d. colder and moister

ANSWER: b

7.   A moist, tropical air mass that is warmer than the surface over which it is moving would be classified:
     a. mTw
     b. cTk
     c. mTk
     d. cPk
     e. mPw

ANSWER: a

8.   The greatest contrast in both temperature and moisture will occur along the boundary separating which air masses?
     a. cP and cT
     b. mP and mT
     c. mP and cT
     d. mT and cP
     e. cT and mT

ANSWER: d

9.   The greatest contrast in both temperature and moisture will occur along the boundary separating which air masses?
     a. cP and cT in summer
     b. mP and mT in winter
     c. cP and mT in summer
     d. cA and mT in winter
     e. mP and mT in summer

ANSWER: d

10. An air mass is characterized by similar properties of ___ and ___ in any horizontal direction.
     a. temperature, pressure
     b. pressure, moisture
     c. winds, moisture
     d. temperature, moisture

ANSWER: d

11. One would expect a cP air mass to be:
     a. cold and dry
     b. cold and moist
     c. warm and dry
     d. warm and moist

ANSWER: a

12. A Texas norther (or blue norther) is most often associated with which air mass?
    a. mT
    b. mP
    c. cA
    d. cT

ANSWER: c

13. When cP air moves into western Washington, western Oregon, and California from the east, the air mass is warmer at the surface than it was originally because:
    a. the sun heats the air
    b. the air sinks, is compressed, and warms
    c. the ocean warms the air
    d. friction with the ground warms the air
    e. latent heat of condensation warms the air as it moves downhill

ANSWER: b

14. Which air mass would show the most dramatic change in both temperature and moisture content as it moves over a large body of very warm water?
    a. cT in summer
    b. cP in winter
    c. mP in winter
    d. mT in summer

ANSWER: b

15. The coldest of all air masses is:
    a. mT
    b. mP
    c. cT
    d. cF
    e. cA

ANSWER: e

16. What type of air mass would be responsible for refreshing cool, dry breezes after a long summer hot spell in the Central Plains?
    a. mP
    b. mT
    c. cP
    d. cT

ANSWER: c

17. Record breaking low temperatures are associated with which air mass?
    a. mT
    b. mP
    c. cP
    d. cT

ANSWER: c

18. Clear sunny days with very cold nights would be associated with what type of air mass?
    a. mP
    b. mT
    c. cP
    d. cT

ANSWER: c

19. Cumuliform cloud development would be most likely in which of the following?
    a. cT air mass moving over a mountain range
    b. cP air mass moving over warm water
    c. mT air mass moving over cold land surface
    d. cT air mass moving over cold water

ANSWER: b

20. Lake-effect snows are best developed around the Great Lakes during:
    a. early spring when moist, tropical air moves over the frozen lakes
    b. late fall and early winter when cold, dry polar air moves over the relatively
       warm water
    c. late fall and early winter when moist, polar air sweeps in from the east
    d. middle winter when the unseasonably warm air mass moves over the cold water

ANSWER: b

21. The lake effect occurs when ___ air mass moves over a ___ body of water.
    a. an mT, cold
    b. an mT, warm
    c. a cP, cold
    d. a cP, warm

ANSWER: d

22. Generally, the greatest lake effect snow fall will be on the ___ shore of the Great Lakes.
    a. northern
    b. southern
    c. eastern
    d. western

ANSWER: c

228

228

23. During the winter, an air mass that moves into coastal sections of Oregon and Washington from the northwest would most likely be:
    a. mP
    b. mT
    c. cP
    d. cT

ANSWER: a

24. Wintertime mP air masses are less common along the Atlantic coast of North America than along the Pacific coast mainly because:
    a. the water is colder along the Pacific coast
    b. the prevailing winds aloft are westerly
    c. the source region for mP air on the Atlantic coast is western Europe
    d. the water is warmer along the Atlantic coast
    e. the land is colder along the Atlantic coast

ANSWER: b

25. The designation for a cool, moist air mass is:
    a. mT
    b. mP
    c. cT
    d. cP

ANSWER: b

26. What type of air mass would be responsible for snow showers on the western slopes of the Rockies?
    a. mT
    b. cP
    c. mP
    d. cA

ANSWER: c

27. What type of air mass would be responsible for persistent cold, damp weather with drizzle along the east coast of North America?
    a. mP
    b. mT
    c. cP
    d. cT
    e. cA

ANSWER: a

28.   A warm, moist air mass that forms over water is called:
        a. cP
        b. mP
        c. mT
        d. wT

ANSWER: c

29.   What type of air mass would be responsible for hot, muggy summer weather in the eastern half of the United States?
        a. mP
        b. mT
        c. cP
        d. cT
        e. cA

ANSWER: b

30.   The air mass with the highest actual water vapor content is:
        a. mT
        b. cT
        c. mP
        d. cP

ANSWER: a

31.   In Southern California, which air mass is mainly responsible for heavy rains, flooding in low-lying valleys, and melting of snow at high elevations?
        a. mT
        b. mP
        c. cP
        d. cA

ANSWER: a

32.   During the spring, which air mass would most likely bring record-breaking high temperatures to the eastern half of the United States?
        a. mT
        b. mP
        c. cP
        d. cT

ANSWER: a

33.   What type of air mass would be responsible for daily afternoon thunderstorms along the Gulf Coast?
       a. mP
       b. mT
       c. cP
       d. cT

ANSWER: b

34.   What type of air mass would be responsible for heavy summer rain showers in southern Arizona?
       a. cP
       b. cT
       c. mP
       d. mT

ANSWER: d

35.   An mT air mass lying above a cold ground surface represents a(an) ___ situation: a. stable
       b. unstable
       c. occluded
       d. stationary

ANSWER: a

36.   What type of air mass would be responsible for summer afternoon thunderstorms along the eastern slopes of the Sierra Nevada mountains in California?
       a. mP
       b. mT
       c. cT
       d. cP
       e. cA

ANSWER: b

37.   Which air mass forms over North America only in summer?
       a. mT
       b. mP
       c. cT
       d. cP

ANSWER: c

38. What type of air mass would be responsible for hot, dry summer weather in southern Arizona?
   a. mP
   b. mT
   c. cP
   d. cT

ANSWER: d

39. A maritime polar air mass that reaches the Pacific Coast is
   a. cool
   b. moist
   c. conditionally unstable
   d. all of the above

ANSWER: d

40. During winter, easterly winds along the front range of the Rocky Mountains provides excellent conditions for
   a. lake-effect snow
   b. maritime tropical air
   c. upslope snow
   d. nor'easters

ANSWER: c

41. Along the boundary between continental polar and maritime tropical air masses, _____ is often found.
   a. a large area of calm (extremely light wind)
   b. intense heat and drought
   c. widespread precipitation and storminess
   d. both a and c

ANSWER: c

42. On a weather map, the transition zone between two air masses with sharply contrasting properties is marked by:
   a. the letter "H"
   b. the words "air mass weather"
   c. a front
   d. the letter "L"

ANSWER: c

43.   The word "frontogenesis" on a weather map would mean that:
       a. a front is in the process of dissipating
       b. one front is about to over take another front
       c. a front is regenerating or strengthening
       d. severe thunderstorms will form along a front

ANSWER: c

44.   Continental tropical air masses are typically found in
       a. northern Mexico
       b. the Gulf of Mexico
       c. the southwestern U.S.
       d. the southeastern U.S.
       e. both a and c

ANSWER: e

45.   Fronts are associated with
       a. low pressure
       b. high pressure

ANSWER: a

46.   The only indication on the station model of past weather conditions is the
       a. temperature
       b. dew point
       c. cloud cover
       d. wind direction
       e. pressure tendency

ANSWER: e

47.   A drylines is
       a. a stalled cold front
       b. a stalled warm front
       c. a dew point front
       d. a boundary marking a strong horizontal change in atmospheric moisture
       e. both c and d

ANSWER: e

48.   An upper-air front involves downward motion of the
       a. tropopause
       b. stratopause
       c. low-level jet
       d. both a and c

ANSWER: a

49.   When comparing an "average" cold front to an "average" warm front, which of the following is not correct?
       a. generally, cold fronts move faster than warm fronts
       b. generally, cold fronts have steeper slopes
       c. generally, precipitation covers a much broader area with a cold front
       d. especially in winter, cumuliform clouds are more often associated with cold fronts

ANSWER: c

50.   Alternating lines of blue and red on a surface weather chart indicate:
       a. a cold front
       b. a warm front
       c. a stationary front
       d. an occluded front

ANSWER: c

51.   A stationary front does not move because:
       a. winds on both sides of the front are calm
       b. the winds blow parallel to the front
       c. the front is between high and low pressure
       d. the winds blow against each other and are of equal strength

ANSWER: b

52.   A true cold front on a weather map is always:
       a. associated with precipitation
       b. associated with a wind shift
       c. followed by drier air
       d. followed by cooler air

ANSWER: d

53.   Which of the following is not correct concerning a cold front?
       a. it marks the position of a trough of low pressure
       b. it marks a zone of shifting winds
       c. it is colored purple on a weather map
       d. it has cold air behind it

ANSWER: c

234

54. On a weather map this front, drawn in blue, represents a region where colder air is replacing warmer air:
    a. warm front
    b. cold front
    c. cold-type occluded front
    d. warm-type occluded front

ANSWER: b

55. Before the passage of a cold front the pressure normally ___, and after the passage of a cold front the pressure normally ___.
    a. drops, drops
    b. drops, rises
    c. rises, rises
    d. rises, drops

ANSWER: b

56. Squall lines most often form ahead of a:
    a. cold front
    b. warm front
    c. cold-type occluded front
    d. warm-type occluded front
    e. stationary front

ANSWER: a

57. In winter, thunderstorms are most likely to form along:
    a. cold fronts
    b. warm fronts
    c. stationary fronts
    d. occluded fronts

ANSWER: a

58. A "back door" cold front moving through New England would most likely have winds shifting from ___ to ___.
    a. easterly, westerly
    b. westerly, northeasterly
    c. southerly, westerly
    d. southeasterly, southwesterly
    e. westerly, southerly

ANSWER: b

59. A "back door" cold front describes which of the following situations?
     a. a cold front moving into Washington state from the Pacific ocean
     b. a cold front moving into the desert southwest from Northern Mexico
     c. a cold front that moves into New England from the Atlantic Ocean
     d. a cold front that moves in a clockwise direction around a low pressure center

ANSWER: c

60. A cold front that moves into New England from the east or northeast is called:
     a. a cold occluded front
     b. an oceanic front
     c. a nor'easter
     d. a "back door" cold front

ANSWER: d

61. Which of the following is not correct concerning a warm front?
     a. it is colored red on a weather map
     b. it has warm air ahead (in advance) of it
     c. in winter it is usually associated with stratiform clouds
     d. it normally moves more slowly than a cold front

ANSWER: b

62. A halo around the sun or moon indicates that rain may be on the way because the halo indicates:
     a. a cold front may be approaching
     b. a warm front may be approaching
     c. a sharp drop in atmospheric pressure
     d. a sudden rise in surface dew point temperature

ANSWER: b

63. A frontal inversion would probably be best observed:
     a. with a warm front in summer
     b. with a stationary front in winter
     c. with a cold front is summer
     d. with a warm front in winter

ANSWER: d

64. In winter, which sequence of clouds would you most likely expect to observe as a warm front with precipitation approaches your location?
     a. cirrus, nimbostratus, altostratus, cumulonimbus
     b. cirrus, cirrostratus, altostratus, nimbostratus
     c. cirrostratus, nimbostratus, altostratus, fog
     d. cirrus, cirrostratus, altostratus, cumulonimbus

ANSWER: b

65. The rising of warm air up and over cold air is called:
    a. overrunning
    b. frontolysis
    c. frontogenesis
    d. occlusion

ANSWER: a

66. At a warm front, the warm air:
    a. rises and cools
    b. rises and warms
    c. sinks and cools
    d. sinks and warms

ANSWER: a

67. During the winter as you travel toward a warm front, the most likely sequence of weather you would experience is:
    a. snow, freezing rain, hail, sleet
    b. rain, snow, sleet, freezing rain
    c. freezing rain, snow, sleet, rain
    d. snow, sleet, freezing rain, rain

ANSWER: d

68. On a weather map where cold air is replacing cool air, what type of front is drawn?
    a. warm front
    b. cold front
    c. warm-type occluded front
    d. cold-type occluded front

ANSWER: d

69. Occluded fronts may form as:
    a. a cold front overtakes a warm front
    b. a warm front overtakes a cold front
    c. a cold front overtakes a squall line
    d. overrunning occurs along a warm front

ANSWER: a

70. Which below is not correct concerning an occluded front?
    a. it is often associated with a broad band of precipitation
    b. it marks a zone of shifting wind
    c. it is colored purple on a surface weather map
    d. at the surface it is always followed by colder air

ANSWER: d

71.   A cold-type occluded front:
        a. has cold surface air ahead of it
        b. has warm surface air behind it
        c. has cold surface air behind it
        d. has cold air rising above warmer air

ANSWER: c

72.   What type of weather front would be responsible for the following weather forecast: "Increasing cloudiness and warm today with the possibility of showers by this evening. Turning much colder tonight. Winds southwesterly becoming gusty and shifting to northwesterly by tonight."
        a. cold front
        b. warm front
        c. cold-type occluded front
        d. stationary front

ANSWER: a

73.   What type of weather front would be responsible for the following weather forecast: "Increasing high cloudiness and cold this morning. Clouds increasing and lowering this afternoon with a chance of snow or rain tonight. Precipitation ending tomorrow morning. Turning much warmer. Winds light easterly today becoming southeasterly tonight and southwesterly tomorrow."
        a. cold front
        b. warm front
        c. stationary front
        d. warm-type occluded front

ANSWER: b

74.   What type of weather front would be responsible for the following weather forecast?
        "Cool today with rain becoming heavy by this afternoon. Slightly warmer tomorrow. Winds southeasterly becoming westerly to northwesterly by tomorrow morning."
        a. cold front
        b. warm front
        c. cold-type occluded front
        d. warm-type occluded front

ANSWER: d

75. What type of weather front would be responsible for the following weather forecast? "Light rain and cold today with temperatures just above freezing. Southeasterly winds shifting to westerly tonight. Turning colder with rain becoming mixed with snow, then changing to snow."
     a. cold front
     b. warm front
     c. cold-type occluded front
     d. warm-type occluded front

ANSWER: c

76. In the *Southern* Hemisphere,
     a. air rises steeply ahead of a warm front, and gradually ahead of a cold front
     b. cold, warm and occluded fronts are associated with high pressure systems
     c. precipitation associated with warm fronts tends to be brief and intense
     d. precipitation associated with warm fronts tends to be gentle and prolonged

ANSWER: d

Essay Exam Questions

1. List the four basic types of air masses. Give an example of where each type could originate and describe how each air mass could affect local weather conditions if it moved into your region.

2. When a warm and cold air mass collide, the warm air is forced upward. Why does this occur?

3. What type of clouds, if any, would you expect to see form when a cP air mass moves across warm water? Would conditions be any different when mT air moves across a cold land surface? Would types of clouds would form in this latter case?

4. Draw side views of a typical warm and cold front. Clearly indicate the temperatures of the separate air masses and show their directions of motion. What types of clouds would you expect to find and where? Where would you expect precipitation to occur?

5. Describe some of the changes in weather conditions (winds, temperature, clouds, precipitation, pressure changes) you would expect to observe as a cold front approaches and passes through your location.

6. How would a warm front, a cold front, and a center of low pressure appear on a surface weather map in the Southern Hemisphere?

7. Explain why the observation of a halo around the moon or sun could forewarn of the arrival of a warm front.

8. Is it possible for a stationary front to produce precipitation? If so, would you expect to find the precipitation on the warm or the cold air side of the front?

9. *Without considering changes in temperature*, how are the weather conditions accompanying a cold frontal passage different from those associated with the passage of a warm front?

10. Explain how, using no meteorological instruments other than your eyes, you could identify the passage of an occluded front.

Chapter 13
Middle Latitude Cyclones

Summary

The structure of middle latitude storm systems and some of the factors that govern their development are covered in this chapter. The chapter begins with a review of the polar front theory of wave cyclones, originally formulated by a group of Norwegian meteorologists after the end of World War I. The vertical structure of middle latitude storms and the influence of upper-level wind patterns on storm development are examined next. Students will see, for example, that diverging motions at upper levels can create a center of low pressure at the surface or cause an existing low to intensify. Upper level winds also determine the direction of movement of surface cyclones and anticyclones. An illustration of cyclogenesis, the development or strengthening of a cyclone, is given in the context of the *nor'easter*.

The chapter explains how fast moving shortwaves can perturb the upper-level longwave flow pattern and produce regions of temperature advection, or baroclinic zones in the atmosphere. The resulting air motions can enhance storm development. Strong upper level divergence is also found at certain positions near the jet streak, a region of strong winds in the core of the jet stream. The concept of vorticity is introduced and the relationship between vorticity, upper level divergence, rising air motions and surface storm development is examined. Vorticity is used to explain why an area of low pressure will often form on the downwind side of a mountain range. Polar lows, strong cyclones occurring poleward of the polar front, are also described.

Material introduced in this chapter is tied together with the so-called 'Storm of the Century' that ravaged the east coast of North America during March, 1993.

Key Terms

polar front theory
frontal wave
incipient cyclone
wave cyclone
open wave
warm sector
triple point
secondary low
"family" of cyclones
cyclogenesis
explosive cyclogenesis
 ("bomb")
convergence
divergence
unstable waves
stable waves
Hatteras low
Alberta Clipper

longwave
Rossby waves
shortwave
retrograde
barotropic (atmosphere)
baroclinic (atmosphere)
baroclinic wave
cold advection
warm advection
theory of developing
 cyclones
wind speed shear
baroclinic instability
differential
 temperature advection
bent back occlusion
cut-off low

vertically stacked
 (storm system)
filling (low)
conveyor belt model
comma cloud
warm conveyor belt
cold conveyor belt
dry conveyor belt
vorticity
earth's vorticity
relative vorticity
absolute vorticity
Coriolis parameter
vorticity maximum
vorticity minimum
lee-side lows
polar lows
arctic front

Teaching Suggestions

1.   The Pacific and Atlantic surface analysis maps will generally show several middle latitude storm storms approaching and leaving the US. Examples of storms in various stages of development can often be seen.

2.   Students will often have difficulty, initially, trying to locate shortwaves on an upper-level chart. Instead of a single chart, show the students a sequence of charts covering a period of 2 or 3 days. Shortwaves will be more apparent as they move and distort different portions of the longwave pattern. Contours on the 500 mb analysis and forecast maps, showing local vorticity maxima and minima, are helpful also.

3.   Discuss the relationship between convergence/divergence and vertical motion using the following analogy. Imagine a coffee can full of marbles, with several small holes drilled in the bottom of the can. If the bottom of the can represents the surface and the top represents 500 mb, and the number of marbles in the can represents the surface air pressure, then how can the surface pressure decrease when surface-level convergence puts more marbles in the can?

Student Projects

1.   Have students track the development and movement of middle latitude storms as they approach and move across the United States. Have students identify associated features on the upper-air charts.

Blue Skies   2.   Use the Weather Forecasting/Forecasting section of the BlueSkies cdrom to examine the relationship between temperature advection, fronts and isobars. Describe this relationship.

242

3.   Use the Weather Forecasting/Forecasting section of the BlueSkies cdrom to identify any large high or low pressure systems in the United States. Do you think these systems are intensifying or decaying? Why?

Answers to Questions for Thought

1.   The storm system began as a frontal wave along the eastern slope of the Rocky Mountains. It developed into an open wave cyclone over the Plains, became occluded over the North Atlantic Ocean, and eventually moved northeastward into Great Britain.

2.   The wave cyclone would dissipate. The air aloft would be converging and descending directly above the surface cyclone.

3.   Cold advection normally occurs to the west of the upper-level trough where the air aloft is also slowly sinking and warming.

4.   Baroclinic conditions exist only in isolated regions of the atmosphere, such as above the polar front, where there are large pressure and temperature gradients.

5.   Baroclinic waves require sharp temperature contrasts that are usually found along frontal boundaries. Such temperature contrasts and fronts are usually not found in the tropics.

6.   On a non-rotating earth, the earth's vorticity would be zero. An air parcel would continue to move in a straight line in the same direction unless acted on by some external force, such as the pressure gradient force.

7.   Pacific storms often re-develop on the eastern slopes of the Sierra Nevada Mountains because as the air moves up and over the mountain barrier, the air's vorticity changes. This creates a cyclonic bending of the air flow on the downwind (leeward) side of the barrier.

8.   The cut-off low would most likely appear as a closed isotherm representing a region of cold air.

9.   The water surface could be warmer than the air above. In this case, the water would warm the air just above and the environmental temperature would decrease rapidly with increasing altitude (steep environmental lapse rate). Development of cumuliform clouds under these conditions would be similar to the Lake Effect.

10.   It means that above 500 mb is the vertical dividing line between convergence and divergence. If divergence is occurring above the 500-mb level, then convergence is occurring below this level. If convergence is occurring above the 500-mb level, then divergence is occurring below.

Answers to Problems and Exercises

2. (a) Absolute vorticity will decrease as the parcel moves upstream and away from Point m.
    (b) $\Delta\zeta_a/\Delta t$ is negative downstream from Point m.
    (c) same region as in (b).

Multiple Choice Exam Questions

1. The polar front theory of a developing wave cyclone was conceived in:
    a. Norway
    b. Great Britain
    c. United States
    d. Germany
    e. Soviet Union

ANSWER: a

2. Which of the scientists below was not one of the meteorologists who helped develop the polar front theory of a developing wave cyclone?
    a. B. Bjerknes
    b. C.G. Rossby
    c. T. Bergeron
    d. H. Solberg
    e. J. Bjerknes

ANSWER: b

3. On a surface weather map that shows an open wave cyclone, the warm sector can be observed:
    a. ahead of an advancing cold front
    b. behind an advancing cold front
    c. ahead of an advancing cold-occluded front
    d. behind an advancing cold-occluded front
    e. ahead of an advancing warm front

ANSWER: a

4. According to the model of the life cycle of a wave cyclone, the storm system is normally most intense:
    a. as a frontal wave
    b. as a stable wave
    c. as an open wave
    d. as a stationary wave
    e. when the system first becomes occluded

ANSWER: e

5.   Which below is not a name given to a large cyclonic storm system that forms in the middle latitudes?
     a. middle latitude cyclone
     b. extratropical cyclone
     c. wave cyclone
     d. anticyclone

ANSWER: d

6.   In the polar front theory of a developing wave cyclone, energy for the storm is usually derived from all but one of the following:
     a. rising of warm air and the sinking of cold air
     b. latent heat of condensation
     c. an increase in surface winds
     d. heat energy stored in the ground

ANSWER: d

7.   Another term for explosive cyclogenesis used by meteorologists is:
     a. lee-side low development
     b. cut-off low cyclogenesis
     c. deepening shortwave
     d. baroclinic development
     e. bomb

ANSWER: e

8.   The development or strengthening of a middle latitude storm system is called:
     a. convergence
     b. divergence
     c. cyclogenesis
     d. frontolysis

ANSWER: c

9.   Which region is not considered to be a region where cyclogenesis often occurs:
     a. eastern slopes of the Rocky Mountains
     b. Atlantic Ocean near Cape Hatteras, North Carolina
     c. California
     d. the Great Basin of the United States
     e. Gulf of Mexico

ANSWER: c

10. A building anticyclone means:
   a. the central pressure is increasing
   b. the anticyclone is moving toward the east coast
   c. separate anticyclones are merging
   d. the anticyclone is causing a middle latitude storm to form

ANSWER: a

11. For cyclogenesis to occur along a frontal wave, the winds aloft directly above the wave should be:
   a. diverging
   b. converging
   c. blowing straight from west to east
   d. increasing in speed uniformly over a broad area

ANSWER: a

12. The piling up of air above a region is called:
   a. thickening
   b. divergence
   c. cyclogenesis
   d. convergence

ANSWER: d

13. If the flow of air into a surface low pressure area is greater than the divergence of air aloft, the surface pressure in the center of the low will:
   a. increase
   b. decrease
   c. remain the same
   d. deepen

ANSWER: a

14. Which of the following is not associated with rising air motions?
   a. overrunning
   b. convergence of air at the surface
   c. convergence of air aloft
   d. divergence of air aloft

ANSWER: c

15. Cyclogenesis is the _____ of a mid-latitude cyclone.
    a. development or strengthening
    b. weakening or dissipation
    c. term for the exact midpoint
    d. none of the above

ANSWER: a

16. A lee-side low forms
    a. over the central Pacific Ocean
    b. near the equator
    c. on the upwind side of a mountain
    d. on the downwind side of a mountain

ANSWER: d

17. Northeasters (or nor'easters) are midlatitude storms commonly found
    a. along the Pacific coast of North America
    b. along the Atlantic coast of North America
    c. along the Gulf coast of North America
    d. both a and b

ANSWER: b

18. A surface low pressure area with a deep upper-level trough to the west will tend to move toward the:
    a. northwest
    b. northeast
    c. southwest
    d. southeast

ANSWER: b

19. When a deep upper-level trough is located to the east of a surface anticyclone, the surface anticyclone will tend to move toward the:
    a. northwest
    b. northeast
    c. southwest
    d. southeast

ANSWER: d

20.   When an upper-level low lies directly above a surface low:
        a. the surface low will probably weaken
        b. thunderstorms will develop
        c. a wave cyclone will begin to form
        d. the pressure of the surface low will decrease
        e. cyclogenesis will occur

ANSWER: a

21.   An upper-level pool of cold air that has broken away from the main flow is called:
        a. a cut-off low
        b. a shortwave
        c. a wave cyclone
        d. a lee-side low

ANSWER: a

22.   For a surface storm system to intensify, the upper-level low (or trough) should be located to the
        _ of the surface low.
        a. north
        b. south
        c. east
        d. west

ANSWER: d

23.   Developing low pressure areas generally have ___ air near the surface and ___ air aloft.
        a. converging, diverging
        b. diverging, converging
        c. converging, converging
        d. diverging, diverging

ANSWER: a

24.   Strong storms that develop over water, poleward of the polar front, are called
        a. nor'easters
        b. upslope lows
        c. lee-side lows
        d. polar lows
        e. none of the above

ANSWER: d

25.   Like hurricanes, polar lows have a clear area in their center.
        a. true
        b. false

ANSWER: a

26. When upper-level divergence of air above a surface low pressure area is stronger than the convergence of surface air, the surface pressure will ___ and the storm itself will ___.
    a. increase, intensify
    b. increase, dissipate
    c. decrease, intensify
    d. decrease, dissipate

ANSWER: c

27. If the outflow of air around a surface high pressure area is greater than the convergence of air aloft, you would observe:
    a. in increase in pressure in the center of the high
    b. movement of the high toward the northeast
    c. a decrease in the central pressure
    d. strengthening in the high

ANSWER: c

28. Longwaves in the middle and upper troposphere usually have lengths on the order of:
    a. tens of kilometers
    b. hundreds of kilometers
    c. thousands of kilometers
    d. millions of kilometers

ANSWER: c

29. An upper-level trough that shows retrograde motion would probably be moving toward the:
    a. west
    b. east
    c. north
    d. south

ANSWER: a

30. Rossby waves are also known as:
    a. stable waves
    b. shortwaves
    c. tidal waves
    d. longwaves

ANSWER: d

31. Atmospheric shortwaves usually move ___ at a speed that is ___ than longwaves.
    a. east to west, faster
    b. west to east, faster
    c. east to west, slower
    d. west to east, slower

ANSWER: b

32. Atmospheric shortwaves usually move ___ than longwaves, and ___ when they move through a longwave ridge.
    a. faster, weaken
    b. faster, strengthen
    c. slower, weaken
    d. slower, strengthen

ANSWER: a

33. A small, moving disturbance imbedded in a longwave is called:
    a. a lee-side low
    b. a wave cyclone
    c. a shortwave
    d. a frontal wave

ANSWER: c

34. On an upper-level chart where the isotherms cross the isobars (or contours) and temperature advection occurs, the atmosphere is called:
    a. barotropic
    b. geostrophic
    c. hydrostatic
    d. baroclinic

ANSWER: d

35. During baroclinic instability:
    a. wave cyclones can intensify into large storm systems
    b. strong wind speed shear exists from the surface up to at least the 500 mb level
    c. rising and descending air motions exist
    d. temperature advection is occurring
    e. all of the above

ANSWER: e

36. Which of the following statements is <u>not</u> correct about vorticity?
    a. the earth's vorticity in the Northern Hemisphere is positive
    b. the earth's vorticity is zero at the poles
    c. air that spins cyclonically possesses positive vorticity
    d. absolute vorticity is the sum of the earth's vorticity and the relative vorticity

ANSWER: b

37. If we assume that the absolute vorticity of flowing air is conserved, air moving northeastward will bend ___ to compensate for the ___ in the earth's vorticity.
    a. anticyclonically, decrease
    b. anticyclonically, increase
    c. cyclonically, increase
    d. cyclonically, decrease

ANSWER: b

38. Lee-side lows are:
    a. storms (extratropical cyclones) that form on the eastern (lee) side of a mountain range
    b. thermal lows that form due to surface heating over deserts
    c. storms (tropical cyclones) that form near the Leeward Islands
    d. low pressure areas that form over oceans, then move onshore of the lee side of the coastline

ANSWER: a

39. The planetary vorticity of an air parcel moving from low toward high latitude in the Northern Hemisphere will:
    a. increase
    b. decrease
    c. remain constant
    d. change from positive to negative

ANSWER: a

40. Vorticity refers to:
    a. the rising and sinking of air along weather fronts
    b. the formation of clouds
    c. the spin of air parcels
    d. the changing of the seasons
    e. the development of a wave cyclone

ANSWER: c

41.   The type of weather system known as a 'mid-latitude cyclone' cannot form over the tropical ocean because
        a. surface temperature contrasts are not large
        b. the ocean surface has a lot of waves
        c. the Coriolis force is weak in the tropics
        d. both (a) and (b)
        e. both (a) and (c)

ANSWER: e

Essay Exam Questions

1.   Describe or illustrate the various phases in the life cycle of a middle latitude storm according to the polar front theory.

2.   Define the term cyclogenesis. List some of the regions in the United States where cyclogenesis is common.

3.   Draw a sketch of a 500 mb chart that clearly shows a trough and a ridge. Where would you expect to find the maximum and minimum values of absolute vorticity? Where would you expect to find converging and diverging wind motions? Below what point on your 500 mb chart would you expect middle latitude storm development to occur?

4.   With the aid of a diagram, show why an intensifying surface low pressure center which is located just east of a deep upper-level trough will often move in a northeasterly direction.

5.   What does the term "shortwave" refer to? Why is it important locate and follow the movements of atmospheric shortwaves? How is this done? How is a shortwave different from a "longwave?"

6.   Describe, in words or with a sketch, a wind flow pattern that will result in upper-level divergence.

7.   Describe some of the ways in which the upper-level wind flow pattern can influence the development and movement of a middle latitude storm system.

8.   List some of the reasons why the polar front theory doesn't apply to storms in the tropics

9.   List some of the factors that helped make the developing wave cyclone of March 13, 1993 the "storm of the century."

10.   When making a weather forecast, which kind of chart is more important: a surface chart or a 500 mb chart?

Chapter 14
Weather Forecasting

Summary

Weather forecasts are an important part of our daily lives. This chapter looks at how weather observations are collected and analyzed, and at the wide variety of different types of weather forecasts that are made using this data.

We see first in the chapter that forecasts are based on worldwide observations of weather conditions made several times a day. This large data set can be processed very quickly using modern computer technology including AWIPS (Advanced Weather Interactive Processing System). Using this technology, meteorologists provide early warning of developing hazardous weather conditions. The specific conditions that warrant issuance of weather alerts are summarized.

The uses of computers and weather satellites in weather forecasting are examined. Computers perform the numerical calculations in mathematical models of the atmosphere. Some of factors which limit the accuracy of numerical weather predictions are listed and discussed, including the chaos theory.

Different methods of predicting future weather including persistence and steady state forecasts, the analogue method, weather type and climatological forecasts are examined. Often one method is appropriate in a short-range forecast, while another method would be employed when formulating a longer range seasonal outlook. A detailed case-study forecast is presented in a section entitled "A Forecast for Six Cities". This section steps the reader through the mechanics of making forecasts for several locations, including consideration of a variety of maps and forecast methods.

A section entitled "Why Forecasts Go Awry and Steps to Improve Them" examines several reasons for inaccurate forecasts, and discusses several possibilities for improvement, including high-resolution weather prediction models and ensemble forecasting techniques.

The chapter concludes with examples of how fairly accurate short term forecasts can be made using data from surface charts and a knowledge of the upper level wind patterns. Students should find these examples particularly instructive as they apply many of the concepts developed in previous chapters.

Key Terms

World Meteorological
  Organization (WMO)
National Meteorological
  Center (NMC)
Weather Service Forecast
  Offices (WSFO)
Weather Service
  Offices (WSO)
Greenwich Mean Time
  (GMT)
weather watch
weather warning
high wind warning
wind advisory
wind chill advisory
flash-flood watch
flash-flood warning
severe thunderstorm
  watch
severe thunderstorm
  warning
tornado watch
tornado warning
snow advisory
winter storm warning

blizzard warning
dense fog advisory
small craft advisory
gale warning
storm warning
hurricane watch
hurricane warning
analysis
numerical weather
  prediction
atmospheric models
grid points
prognostic chart (prog)
medium-range forecast
AFOS (Automation of
  Field Services and
  Operations)
AWIPS (Advanced
  Weather Interactive
  Processing System)
ASOS (Automated
  Surface Observing
  Systems)
meteogram
soundings

index
wind profilers
chaos
ensemble forecasting
persistence forecast
steady-state (trend)
  forecast
nowcasting
analogue method
pattern recognition
weather types
long-range weather
  forecasting
extended weather
  forecasts
climatological forecast
probability forecast
teleconnections
Book of Signs
veering wind
backing wind
pressure tendency
isallobars
omega high
comma cloud

Teaching Suggestions

1.   The person who presents the local television weather report may be willing to come and speak to the class. Or, in the case of a small class, it might be possible to arrange a visit to the studio to see a weather broadcast. A small class could also, in some cities, visit the local weather service office.

2.   Using the present surface and upper air charts, together with prognostic charts, make a 12, 24 and 36 hour forecast in class. Use a variety of forecast methods or let the students decide what type of forecast method they think would be most suitable. If the upper level forecast charts do not show a large change during the forecast period, a persistence forecast might be most appropriate. The steady state method could be used in the case of an approaching storm system or front. Some storms might resemble events from earlier in the semester in which case an analogue type forecast could be made. Define criteria that will allow verification of the various forecasts.

Student Projects

1.   Provide students with surface and upper level charts and have them predict the future motion of a middle latitude storm system using the methods listed in the text (past motion, upper-level winds, winds

in the warm sector, and movement toward the region of maximum pressure decrease). Have the students compare their predicted location with the actual location at the end of the forecast period.

Using their predictions of storm positions, students could issue specific forecasts for cities that the storm is likely to affect. This exercise could be patterned after the example in the text.

2.   Have students compare the actual year-to-date average weather conditions (mean temperatures and precipitation amounts) with climatological averages for their region. Prepare a climatological forecast for the remainder of the semester or the year.

Students might try to forecast first or last frost dates, first snow fall, first day over 100 °F, or some other event using climatological data. They might modify their forecast by noting whether conditions thus far during the year have been above or below the climatological average. Can they account for any departure from average conditions?

 3.   Use the Weather Forecasting/Forecasting section of the BlueSkies cdrom to examine the relationship between winds, temperature and pressure in the region of a midlatitude cyclone. On which side of the circulation do you find warmer temperatures? On which side do you find colder temperatures? Describe the relationship between temperature advection and the pressure pattern.

 4.   Use the Weather Forecasting/Forecasting section of the BlueSkies cdrom to examine the relationship between winds, temperature and pressure in the region of a midlatitude anticyclone. On which side of the circulation do you find warmer temperatures? On which side do you find colder temperatures? Describe the relationship between temperature advection and the pressure pattern.

Answers to Questions for Thought

2.   Chances are that only one Christmas out of ten will be a white one. If last Christmas turned out to be white, then based on climatology, next Christmas should be non-white. If you forecasted a non-white Christmas and this forecast turned out to be correct, then your forecast was accurate but you showed no skill because your forecast was not better than climatology.

3.   A persistence forecast would call for warm and rain. We would <u>not</u> expect the forecast to be correct because the weather will change dramatically as the cold front passes. A forecast using the steady state method: "Warm and rain, turning colder with rain changing to snow."

4.   A reduction in grid spacing causes a large increase in grid points. For example, in a 100 km x 100 km area, there are 100 grid points if 10 km x 10 km grid spacing is used. If 1 km x 1 km spacing is used, the number of grid points increases to 10,000. Since a set of equations must be solved at each grid point for each model time step, a reduction in grid spacing creates a great challenge for numerical weather prediction computers.

5.   The computer models used in weather forecasting are "sensitive" to the data initially fed into

them. For example, small unpredictable atmospheric fluctuations (chaos), and small errors in the data usually amplify with time as the computer projects the weather days into the future. These small initial imperfections often dominate after several days, causing the forecast chart to show little resemblance to the behavior of the real atmosphere.

6.   Warmer. Current air is from the Canadian Arctic and therefore cold, whereas tomorrow's air will likely originate over the Pacific Ocean west of the Rockies.

Answers to Problems and Exercises

2.   (a) Good chance of precipitation - western third of the nation.
        (b) Above seasonal temperatures - from the Great Lakes to the Gulf Coast.
        (c) Below seasonal temperatures - southwestern United States and New England.
        (d) Generally dry weather - Great Lakes to the Gulf Coast.

Multiple Choice Exam Questions

1.   A weather warning indicates that:
        a. the atmospheric conditions are favorable for hazardous weather over a particular region
        b. hazardous weather is either imminent or occurring within the forecast area
        c. hazardous weather is likely to occur within the forecast area during the next 24 hours
        d. hazardous weather is frequently observed in a particular region

ANSWER: b

2.   A weather watch would probably be issued for which of the following conditions?
        a. there is a chance for tornadoes tomorrow
        b. presently, extremely high winds are occurring at mountain summits
        c. a tornado has been sighted at the outskirts of town
        d. heavy snow has been falling over the forecast area

ANSWER: a

3.   An analysis is:
        a. a forecast chart that shows the atmosphere at some future time
        b. a forecast chart that compares past weather maps with those of the present
        c. a surface or upper-level chart that interprets the present weather patterns
        d. a forecast method used in long range weather prediction
        e. a method used to determine skill in predicting the weather

ANSWER: c

4. Weather observations made by most countries of the world are shared and distributed globally by international agencies.
    a. true
    b. false

ANSWER: a

5. A forecast of an extended period of dry weather would be made for a region beneath:
    a. an upper-level trough
    b. the polar jet stream
    c. a cold pool of air aloft
    d. an upper-level ridge
    e. a shortwave trough

ANSWER: d

6. Suppose it is warm and raining, and a cold front is moving toward your location. Directly behind the cold front it is cold and snowing. Still further behind the front the weather is cold and clearing. If the front is scheduled to pass your area in 6 hours, a persistence forecast for your area for 12 hours from now would be:
    a. cold and snowing
    b. cold and clearing
    c. cold and cloudy
    d. warm and raining
    e. not enough information on which to base a forecast

ANSWER: d

7. The least accurate forecast method of predicting the weather two days into the future during changeable weather conditions is usually the:
    a. trend method
    b. persistence forecast
    c. analogue method
    d. prediction by weather types
    e. numerical weather prediction

ANSWER: b

8. Weather forecast that predicts that the future weather will be the same as the present weather is called:
    a. a steady-state forecast
    b. the trend method
    c. a persistence forecast
    d. the analogue method
    e. an extended weather forecast

ANSWER: c

9.   Which forecasting method assumes that weather systems will move in the same direction and at the same speed as they have been moving?
      a. persistence forecast
      b. probability forecast
      c. weather type forecast
      d. climatological forecast
      e. steady state (trend) forecast

ANSWER: e

10.   A persistence forecast could be quite accurate when:
      a. a frontal system approaches your location at constant speed
      b. you are positioned in the middle of a large, stationary air mass
      c. the weather has been unusually cold for several days
      d. upper level winds blow straight from west to east

ANSWER: b

11.   A weather forecast for the immediate future that employs the trend method is called:
      a. nowcasting
      b. extrapolation
      c. first order forecast
      d. linear forecast

ANSWER: a

12.   By examining a surface map, the movement of a surface low pressure area can be predicted based upon the:
      a. orientation of the isobars in the warm sector
      b. region of greatest pressure decrease
      c. movement during the previous 6 hours
      d. all of the above

ANSWER: d

13.   Predicting the weather by weather types employs which forecasting method?
      a. probability
      b. steady-state
      c. analogue
      d. persistence
      e. guess

ANSWER: c

14.   A forecast method that compares past weather maps and weather patterns to those of the present is:
       a. persistence forecasting
       b. the analogue method
       c. the trend method
       d. nowcasting

ANSWER: b

15.   The forecasting technique that produces several versions of a forecast model, each beginning with slightly different weather information to reflect errors in the measurements, is called:
       a. climatology forecasting
       b. redundancy analysis
       c. persistence forecasting
       d. ensemble forecasting
       e. probability forecasting

ANSWER: d

16.   Suppose that where you live the middle of January is typically several degrees warmer than the rest of the month. If you forecast this "January thaw" for the middle of next January, you would have made a:
       a. forecast based on the analogue method
       b. persistence forecast
       c. forecast based on weather types
       d. probability forecast
       e. climatological forecast

ANSWER: e

17.   A probability forecast that calls for a "40 percent chance of rain" means that:
       a. there is a 40 percent chance that it will not rain within the forecast area
       b. there is a 40 percent chance that any random place in the forecast area will
          receive measurable rain
       c. it will rain on 40 percent of the forecast area
       d. it will rain during 40 percent of the time over the forecast area

ANSWER: b

18. Which of the following is presently a problem with modern-day weather predictions?
    a. computer forecast models make assumptions about the atmosphere that are not always correct
    b. there are regions of the world where only space observations are available
    c. computer models do not always adequately interpret the surface's influence on the weather
    d. the distance between grid points on some models is too large to pick up smaller-scale weather features such as thunderstorms
    e. all of the above

ANSWER: e

19. An accurate forecast:
    a. always shows skill
    b. may or may not show skill
    c. never shows skill
    d. requires complex computer equipment

ANSWER: b

20. For a forecast to show skill it must:
    a. be better than one based on persistence or climatology
    b. be accurate to within 2 °C of the predicted temperature
    c. be accurate for over more than 90% of the forecast area
    d. use the analogue method of forecasting
    e. use a probability

ANSWER: a

21. A prog is:
    a. a chart that interprets the current state of the atmosphere
    b. an instrument that draws lines on an upper-level chart
    c. a new method of forecasting the weather
    d. a forecast chart that shows the atmosphere at some future time
    e. another name for a probability forecast

ANSWER: d

22. The forecasting of weather by a computer is known as:
    a. weather type forecasting
    b. climatology forecasting
    c. extended weather forecasting
    d. analogue prediction
    e. numerical weather prediction

ANSWER: e

23. The greatest improvement in forecasting skill during the past 30 years has been made in forecasting:
    a. snowstorms along the eastern seaboard
    b. the development of middle latitude cyclones in the Pacific Ocean
    c. severe storm warnings for hurricanes and tornadoes
    d. the movement of warm and cold fronts
    e. maximum and minimum temperature 6 to 10 days into the future

ANSWER: c

24. The system designed to replace the aging AFOS system is the:
    a. NEXRAD system
    b. ASOS system
    c. AWIPS system
    d. WMO system
    e. NMC system

ANSWER: c

25. A wind that changes direction in a counterclockwise sense with increasing height is called a(an):
    a. backing wind
    b. Ekman spiral
    c. meridional wind
    d. veering wind

ANSWER: a

26. Warm advection is most likely to occur:
    a. in the center of a cut-off low
    b. from the surface up to the 500 mb level ahead of an advancing warm front
    c. behind a cold front
    d. where the winds back with height
    e. on the western side of a shortwave trough at the 500 mb level

ANSWER: b

27. The National Weather Service system of watches, warnings and advisories generally does not address weather conditions which limit visibility.
    a. true
    b. false

ANSWER: b

28. Suppose there are two cloud layers above you. The lower cloud layer is moving from a westerly direction, while the higher cloud layer is moving from a northwesterly direction. From this observation you conclude that the wind is ___ with height and ___ advection is occurring between the cloud layers.
   a. backing, cold
   b. backing, warm
   c. veering, cold
   d. veering, warm

ANSWER: d

29. When making a forecast, a meteorologist typically examines
   a. surface charts
   b. upper-air charts
   c. radar displays
   d. satellite imagery
   e. all of the above

ANSWER: e

30. Modern computer forecasting models have increasingly smaller grid spacing. This presents which of the following *problems*?
   a. mesoscale weather features can be predicted
   b. small-scale weather features can be resolved by the models
   c. much more computations are needed
   d. none of the above

ANSWER: c

31. *Very short-range forecasts* often utilize which of the following forecast methods?
   a. steady-state or trend method
   b. pattern recognition method
   c. ensemble method
   d. both a and c
   e. both a and b

ANSWER: e

32. Lines connecting points of equal <u>pressure change</u> are called:
   a. isobars
   b. isograds
   c. contours
   d. isotherms
   e. isallobars

ANSWER: e

33. Which below is <u>not</u> a condition associated with an omega high?
    a. it is in the shape of the Greek letter "omega"
    b. it is known as a "blocking high"
    c. it forms when the flow aloft is zonal
    d. it tends to persist in the same geographic area for many days

ANSWER: c

34. A comma cloud is:
    a. an organized band of clouds that looks like a comma on a satellite photograph
    b. a large cumulonimbus cloud whose anvil is in the shape of a comma
    c. a high wispy cirrus cloud that takes on the shape of a comma
    d. the spiraling arms of a hurricane
    e. a band of high cirrus clouds blown by the jet stream

ANSWER: a

35. If you wanted to make a persistence forecast of minimum and maximum temperatures for a particular city, which type of chart would be most helpful?
    a. surface chart
    b. meteogram
    c. 500 mb chart
    d. adiabatic chart
    e. Doppler radar display

ANSWER: b

Essay Exam Questions

1. List some of the ways that you can predict the future movement of surface middle latitude storms.

2. Describe four different types of weather forecasting methods and give an example of each.

3. List some of the factors that affect the accuracy of atmospheric models.

4. With all other factors being equal, will a cloudy day have a higher or lower maximum temperature then a clear day? How about the minimum temperature during a cloudy vs. clear night? Explain.

5. Describe the data and tools that a meteorologist assembles prior to making a weather forecast.

6. What type of forecast method do you think would be most appropriate for a short range (6 to 12 hours), a medium range (24 to 48 hours), a long range (5 to 10 days), or a seasonal (3 month) forecast?

7. About how far apart do you think weather observing stations should be in order to accurately depict a middle latitude storm and to be able to forecast the storm's movement?

8. What local signs would you look for to predict the approach of a low pressure center or a weather front?

9. Assuming you know last night's minimum temperature, what weather information would you use to predict tonight's minimum temperature?

10. Explain how observed teleconnection patterns can help in the preparation of a seasonal weather forecast.

Chapter 15
Thunderstorms and Tornadoes

Summary

Thunderstorms are among the most spectacular and destructive of weather phenomena. This chapter begins by describing the growth and development of these convective storms, from common ordinary thunderstorms (formerly called air mass thunderstorms) to the particular conditions that lead to severe storm formation. Severe thunderstorms are actually a class of storms that include multicell and supercell storms. Large, organized systems of thunderstorms, such as squall lines and mesoscale convective systems, are described here also. Many of the physical characteristics and some of the hazards associated with thunderstorms, such as gust fronts and roll clouds, microbursts, and flash floods are explained and illustrated.

Gust fronts and microbursts, along with their associated phenomena of outflow boundaries, downbursts and macrobursts, are responsible for many incidents of damage to aircraft. These phenomena, along with the technology in place to detect them, are described in some detail. Additional sections on dryline thunderstorms and flash floods, as well as thunderstorm distribution, round out the description of these fascinating storms.

The chapter continues by reviewing our current understanding of thunderstorm electrification mechanisms and summarizing the complex sequence of events that occurs during a cloud-to-ground lightning discharge. Lightning safety rules and basic techniques used for lightning protection are given. The interesting phenomena of sprites and jets, luminous discharges occurring above thunderstorm tops, are presented in a focus section.

Tornadoes are discussed in the final portion of the chapter. Tornadoes occur more frequently in the United States than in any other country in the world. The chapter examines how, when, and where tornadoes form in the US, and explains why tornadoes can be so destructive. We also look at Doppler radar, a remote-sensing technology that is now routinely used to study tornado-producing thunderstorms, and to detect and warn of tornado occurrence. The chapter concludes with a brief discussion of waterspouts.

Key Terms

ordinary thunderstorms
cumulus stage
entrainment
downdraft
mature thunderstorm
dissipating stage
multicell storms
frontal thunderstorms
orographic thunderstorms
severe thunderstorms
tilted updraft
overshooting
gust front
mesohigh
shelf cloud
arcus cloud
roll cloud
downburst
microburst
macroburst
wind shear
LLWAS (Low-level Wind
  Shear Alert System)
JAWS (Joint Airport
  Weather Studies)
NIMROD (Northern
  Illinois Meteorological
  Research on Downbursts)
virga
derecho
supercell storm
squall line
gravity waves

pre-frontal squall-
  line thunderstorm
ordinary squall lines
dryline
dew-point front
flash floods
Mesoscale Convective
  Complexes (MCCs)
Mesoscale Convective
  Systems (MCSs)
lightning
thunder
sonic boom
cloud-to-ground
  lightning
stepped leader
return stroke
dart leader
forked lightning
ribbon lightning
bead lightning
ball lightning
heat lightning
sheet lightning
corona discharge
St. Elmo's Fire
fulgurite
lightning direction-
  finder
sferics
lightning mapper
  sensor
tornadoes

twisters
cyclones
funnel cloud
dust whirl stage
organizing stage
mature stage
shrinking stage
decay stage
tornado families
tornado outbreak
"tornado belt"
multi-vortex tornadoes
suction vortices
Fujita scale
dry tongue
convective instability
mesocyclone
wall cloud
tornado watch
tornado warning
hook echo
Doppler radar
Doppler shift
tornado vortex
  signature (TVS)
Doppler lidar
NEXRAD (Next Generation
  Weather Radar)
waterspout
tornadic waterspout
"fair weather"
  waterspouts

Teaching Suggestions

1.   The growth and decay of an ordinary (air mass) thunderstorm is illustrated well on **time-lapse video photography**. Unfortunately, home video equipment does not usually have a time-lapse filming capability. It may be possible to borrow suitable equipment from a university audio visual department or perhaps a local television station would be willing to record local thunderstorm activity.

2.   Spectacular video footage of tornadoes, often obtained by amateur photographers using home video equipment, is available for purchase from several sources.

Weatherwise is a good source of still photographs of tornadoes, thunderstorms and lightning. See, especially, the winning entries from the annual weather photography contest and weather summary.

Student Projects

1.   Depending on location and time of the year, students might use data from morning surface and upper-level charts to prepare forecasts of thunderstorm activity for the coming afternoon or evening. Let the students decide which of the various weather elements or simple stability indices might be appropriate. Students should validate their forecast and attempt to improve forecast accuracy.

2.   Thunderstorms and lightning are good choices for student photography projects. Still photographs of a developing thunderstorm taken at intervals of 5 or 10 minutes will often illustrate how rapidly cumulonimbus clouds can develop. For a close storm, it will probably be difficult to fit the entire cloud into the camera's field of view unless a short focal length, wide-angle lens, is available.

If nighttime thunderstorm activity is frequent, students can attempt to photograph the cloud-to-ground lightning. Depending on the height of the cloud base and the focal length of the camera lens, the camera will probably need to be located several kilometers away from the thunderstorm and out of the rain. Students should operate from a protected location if the lightning activity is closer than this. The camera should be mounted on a tripod and exposures should be made using the B (bulb) setting and a shutter release cable. Keep the shutter open until a lightning flash is seen. Test exposures can be made to determine when background light will begin to overexpose the film. Close lightning discharges are sufficiently bright, that low speed film (ASA 25 or 64) gives satisfactory results.

By slowly panning the camera, it might be possible to resolve the separate strokes in a lightning flash (see, for example, J. Hendry Jr., "Panning for Lightning," *Weatherwise, 45,* 19, 1993.)

3.   Ask students to describe a close encounter with a tornado that they or a family member may have had.

 4.   Use the Weather Forecasting/Forecasting activity on the BlueSkies cdrom to identify areas where you think thunderstorms may occur. Then use the Atmospheric Basics/Layers of the Atmosphere section of the cdrom to examine soundings for these areas. Do the temperature and moisture characteristics in the soundings support your prediction for thunderstorms? Why or why not?

 5.   Using the Weather Forecasting/Forecasting activity on the BlueSkies cdrom, examine the current pattern of precipitation in the United States (click the "Radar" overlay). Which precipitation areas are likely due to thunderstorms (as opposed to stratiform precipitation)? Explain your answer.

Answers to Questions for Thought

1. The sinking air increases in strength in the middle and lower portion of the cloud, causing the liquid cloud particles to evaporate quickly. In the upper portion of the cloud, where ice particles prevail and downward motions are slight, the cloud particles survive for a long time.

2. As drier air is drawn into the cloud, some of the falling raindrops evaporate. This cools the sinking air. Actually, the sinking air beneath the cloud base probably warms, but at the moist adiabatic rate, not the dry adiabatic rate. Consequently, when the descending air reaches the surface it is cooler than the air it replaces.

3. The conditions that lead to the lifting of warm, humid air and the generation of a squall line usually occur in the warm sector ahead of an advancing cold front. Behind a cold front the air motions are usually downward, and the air is cooler and drier.

4. A "right moving" thunderstorm is one that moves to the right of the wind.

5. A lightning flash often consists of a series of very rapid strokes that strike the same spot many times.

6. The wisest thing to do is to seek cover inside. If it is not possible to go inside, then crouch. This will minimize the person's contact area with the ground and a crouching person will not be acting as a protruding object.

7. Most tornadoes rotate counterclockwise. When facing an on-rushing tornado it would probably be wisest to run to your right and lie down in a depression. On this side of the twister the wind speed should be slightly lower.

8. The rapid decrease in pressure causes a lowering of the condensation level, and so the tornado cloud forms at successively lower altitudes.

9. On the Fujita scale, most waterspouts would be classified as F_0.

Answers to Problems and Exercises

2. Thunderstorms would most likely be forming at point 3, along the dry line separating the warm, dry air from the warm, humid air.

4. 2 miles.

Multiple Choice Exam Questions

1. All thunderstorms require:
    a. hot, humid air
    b. divergence of the air aloft
    c. lifting along some barrier such as a mountain or front
    d. surface heating
    e. rising air

ANSWER: e

2. The initial stage of an ordinary thunderstorm is the:
    a. mature stage
    b. dissipating stage
    c. cumulus stage
    d. multicell stage

ANSWER: c

3. Ordinary thunderstorms only last about one hour and begin to dissipate when:
    a. lightning neutralizes all the electrical charge in the cloud
    b. when all the precipitation particles in the cloud turn to ice
    c. when the downdraft spreads throughout the cloud and cuts off the updraft
    d. when solar heating at the ground begins to decrease

ANSWER: c

4. An ordinary thunderstorm is a:
    a. thunderstorm that does not produce lightning or thunder
    b. thunderstorm that has a tilted updraft and downdraft
    c. scattered or isolated storm that is not severe
    d. thunderstorm that does not produce hail

ANSWER: c

5. Downdrafts spread throughout a thunderstorm during the ___ stage.
    a. cumulus
    b. dissipating
    c. precipitating
    d. developing

ANSWER: b

6.   An ordinary thunderstorm is most intense during the ___ stage.
       a. mature
       b. multicell
       c. cumulus
       d. dissipating

ANSWER: a

7.   The most likely time for an ordinary thunderstorm to form is:
       a. just after sunrise
       b. just before sunrise
       c. around midnight
       d. late afternoon
       e. at noon

ANSWER: d

8.   Severe thunderstorms are different from ordinary thunderstorms in that severe thunderstorms:
       a. contain thunder and lightning
       b. have an anvil
       c. contain hail
       d. have a strong updraft and downdraft
       e. have a tilted updraft in the mature stage

ANSWER: e

9.   A supercell storm is:
       a. a thunderstorm that produces several tornadoes
       b. a thunderstorm that produces a category F-5 tornado
       c. a thunderstorm with an extremely large downburst
       d. an enormous thunderstorm that lasts for several hours

ANSWER: d

10.   Severe thunderstorms are capable of producing:
       a. large hail
       b. tornadoes
       c. flash floods
       d. all of the above

ANSWER: d

11.   In a severe thunderstorm, hail may:
      a. fall from the base of the anvil
      b. fall from the bottom of the cloud
      c. be tossed out the side of the cloud
      d. all of the above

ANSWER: d

12.   A Mesoscale Convective Complex is actually:
      a. a rapidly rotating tornado cyclone inside a massive thunderstorm
      b. another name for a suction vortex
      c. a complex display of lightning from distant thunderstorms
      d. a family of tornadoes that do a great deal of damage
      e. individual thunderstorms that grow into a large, long-lasting weather system

ANSWER: e

13.   Thunderstorms which produce tornadoes:
      a. have very little cloud-to-ground lightning
      b. have updraft velocities that exceed 100 miles per hour
      c. have rotating updrafts
      d. will not produce hail

ANSWER: c

14.   The downdraft in an ordinary thunderstorm is created mainly by:
      a. the melting of snow in the anvil
      b. electrical attraction between the cloud and ground
      c. the release of latent heat as water in the cloud freezes
      d. evaporating raindrops that make the air cold and heavy
      e. upper level wind motions

ANSWER: d

15.   The cloud that forms along the leading edge of a gust front is called:
      a. an anvil cloud
      b. a roll cloud
      c. a mammatus cloud
      d. a wedge cloud

ANSWER: b

16.   The leading edge of a thunderstorm's cold downdraft is known as a:
        a. downburst
        b. squall line
        c. gust front
        d. dry line
        e. microburst

ANSWER: c

17.   The small area of high pressure created by the cold, heavy air of a thunderstorm downdraft is called:
        a. an anticyclone
        b. a microburst
        c. a mesohigh
        d. a gust front
        e. a mesocyclone

ANSWER: c

18.   Which of the following would you not expect to observe during the passage of a gust front?
        a. gusty winds
        b. rising surface pressures
        c. increase in temperatures
        d. wind shift

ANSWER: c

19.   A small thunderstorm cloud with virga falling out of its base and blowing dust at the ground could warn of a severe hazard to an airplane because:
        a. this could be the first indication of a tornado
        b. it is likely that hail will soon begin to fall
        c. this could indicate an intense downdraft or microburst
        d. the airplane could be struck by lightning

ANSWER: c

20.   The main difference between a downburst and a microburst is:
        a. duration
        b. strength
        c. size
        d. altitude

ANSWER: c

21. The wind shear associated with several major airline crashes is believed to have been caused by:
    a. microbursts
    b. dry lines
    c. the jet stream
    d. mesocyclones

ANSWER: a

22. A relatively narrow downburst, less than 4 kilometers wide, is called:
    a. a microburst
    b. a funnel cloud
    c. a rain shaft
    d. a narrow burst
    e. a mesocyclone

ANSWER: a

23. A group of thunderstorms which develop in a line one next to the other, each in a different stage of development, are called:
    a. ordinary thunderstorms
    b. a thunderstorm cluster
    c. a multicell thunderstorm
    d. mature thunderstorms

ANSWER: c

24. During the summer, what conditions prevail near the surface over the Great Plains that help a hailstone survive as ice all the way to the ground?
    a. a warm, shallow layer of dry air
    b. a warm, thick layer of moist air
    c. a cold, shallow layer of dry air
    d. a cold, thick layer of moist air
    e. a warm, shallow layer of moist air

ANSWER: a

25. Squall lines generally do not form:
    a. behind a cold front
    b. when the air aloft develops waves downwind from a cold front
    c. along a dry line
    d. in the warm sector where warm, dry air meets warm, humid air
    e. ahead of an advancing cold front

ANSWER: a

26. A line of thunderstorms that forms ahead of an advancing cold front is called a:
a. roll cloud
b. squall line
c. wall cloud
d. gust front
e. dry line

ANSWER: b

27. On a surface weather map, this marks the boundary where a warm, dry air mass encounters a warm, moist air mass:
a. gust front
b. dry line
c. storm front
d. wall cloud

ANSWER: b

28. Most squall line thunderstorms form:
a. in advance of a cold front
b. along a cold front
c. behind a cold front
d. in advance of a warm front
e. along an occluded front

ANSWER: a

29. A HP supercell differs from a LP supercell based on the storm's
a. maximum wind gust
b. maximum hail diameter
c. precipitation amount
d. all of the above

ANSWER: c

30. Thunderstorms that move to the right of the winds aloft tend to:
a. form along warm fronts
b. cut off the supply of humid air to thunderstorms to the north of them
c. form in the northern United States and Canada where the Coriolis force is greatest
d. form near mountains where upper-level winds develop into waves
e. form over water due to the reduced friction there

ANSWER: b

274

31. Many flash floods, including those that occurred in Rapid City, South Dakota, and in Colorado's Big Thompson Canyon, are the result of thunderstorms that:
a. contain no lightning
b. form in a dry air mass
c. move slowly
d. have weak or non-existent downdrafts

ANSWER: c

32. The region in the United States with the greatest annual frequency of hailstones is:
a. Florida
b. the Mississippi Valley
c. the Pacific Northwest
d. the western Great Plains
e. Texas

ANSWER: d

33. Which figure comes closest to the estimated number of thunderstorms that occur each year throughout the world?
a. 2,000
b. 40,000
c. 100,000
d. 600,000
e. 14,000,000

ANSWER: e

34. The greatest annual number of thunderstorms in the United States occurs in:
a. the Ohio valley
b. the Central Plains
c. the desert southwest
d. Florida
e. Texas

ANSWER: d

35. In the United States, *dryline thunderstorms* are most common
a. in the Rocky Mountains
b. in the desert southwest
c. in the Great Plains
d. in California
e. in Florida

ANSWER: c

36.   A sonic boom:
        a. is a form of lightning
        b. occurs with lightning
        c. is produced by static electricity in the atmosphere
        d. can be produced by an approaching tornado
        e. is caused when an aircraft flies faster than the speed of sound

ANSWER: e

37.   A discharge of electricity from or within a thunderstorm is called:
        a. static electricity
        b. lightning
        c. a downburst
        d. St. Elmo's fire
        e. an atmospheric arc

ANSWER: b

38.   The radio waves produced by lightning are called:
        a. sferics
        b. St. Elmo's fire
        c. thunder
        d. sonic boom
        e. ball lightning

ANSWER: a

39.   Lightning may occur:
        a. within a cloud
        b. from a cloud to the ground
        c. from one cloud to another cloud
        d. all of the above

ANSWER: d

40.   Lightning discharges within a cloud occur ___ cloud-to-ground lightning.
        a. more frequently than
        b. less frequently than
        c. about as frequently as
        d. lightning cannot remain in the cloud, it must strike an object on the ground

ANSWER: a

41.	In cloud-to-ground lightning, the stepped leader travels ___ and the return stroke travels ___ .
	a. upward, upward
	b. upward, downward
	c. downward, upward
	d. downward, downward

ANSWER: c

42.	A second surge of electrons that proceeds from the base of a cloud toward the ground during cloud-to-ground lightning is called a:
	a. return stroke
	b. stepped leader
	c. dart leader
	d. downstroke
	e. subsequent stroke

ANSWER: c

43.	The bluish halo that may appear above pointed objects underneath a thunderstorm is called:
	a. heat lightning
	b. fluorescence
	c. St. Elmo's fire
	d. sheet lightning

ANSWER: c

44.	St. Elmo's fire is most likely to occur:
	a. over a dry, grassy field
	b. over a thick, moist swamp
	c. near the base of a tree
	d. over a plowed, moist field
	e. at the top of a tall, dead tree

ANSWER: e

45.	Distant lightning that is so far away you cannot hear the thunder is called:
	a. sheet lightning
	b. heat lightning
	c. false lightning
	d. St. Elmo's fire
	e. auroral lightning

ANSWER: b

46.   Electrons:
        a. are negatively charged
        b. are positively charged
        c. carry no charge
        d. can carry either positive or negative charge

ANSWER: a

47.   The upper part of a thunderstorm cloud is normally ___ charged, and the middle and lower parts
        are ___ charged.
        a. negatively, negatively
        b. negatively, positively
        c. positively, positively
        d. positively, negatively

ANSWER: d

48.   The electrification of a cumulonimbus cloud appears to involve:
        a. graupel or hailstones
        b. supercooled cloud droplets
        c. ice crystals
        d. all of the above

ANSWER: d

49.   An important principle in the electrification of a cumulonimbus cloud is that:
        a. raindrops are always positively charged
        b. supercooled water is always negatively charged
        c. ice crystals are always negatively charged
        d. there is a net transfer of positive ions from a warmer object to a colder object

ANSWER: d

50.   A cloud-to-ground lightning discharge will sometimes appear to flicker. This is because:
        a. you are able to see the separate steps of the stepped leader
        b. you are able to distinguish separate return strokes
        c. the bright light causes you to blink
        d. of refraction caused by turbulent thunderstorm winds

ANSWER: b

51.   What would be the proper sequence of events in a lightning flash?
        a. stepped leader, dart leader, return stroke, return stroke
        b. return stroke, stepped leader, return stroke, dart leader
        c. dart leader, return stroke, stepped leader, return stroke
        d. stepped leader, return stroke, dart leader, return stroke

ANSWER: d

52. You a generally safe inside an automobile during a lighting storm because:
    a. the car's radio antenna will act as a lightning rod
    b. the rubber tires insulate you from the ground
    c. metal cars do not become electrically charged
    d. the metal car body will carry the lightning current around the passengers inside

ANSWER: d

53. Which of the following is the most accurate description of the principle of a lightning rod?
    a. the lightning rod acts to discharge the thunderstorm
    b. the lightning rod intercepts the lightning and safely carries the lightning current around the object it protects
    c. lightning rods have been used since the 1700s, but the principle of their operation is not known.
    d. positive charge induced in the lightning rod repels the negative charge in an approaching step leader

ANSWER: b

54. Thunder is caused by:
    a. the collision between two thunderstorms with opposite electrical charge
    b. the rapid heating of air surrounding a lightning channel
    c. the explosion that occurs when + and - charge collide and neutralize each other
    d. turbulent wind motions inside the thunderstorm

ANSWER: b

55. If you see a lightning stroke and then, 15 seconds later, hear the thunder, the lightning is about ____ miles away.
    a. 45
    b. 15
    c. 5
    d. 3

ANSWER: d

56. Thunder will not occur:
    a. without lightning
    b. in wintertime thunderstorms
    c. in thunderstorms over the ocean
    d. when a thunderstorm is producing precipitation

ANSWER: a

57. When caught in a thunderstorm in an open field, the best thing to do is to:
a. run for cover under the nearest tree
b. lie down flat on the ground
c. crouch down as low as possible while minimizing contact with the ground
d. remove all metallic objects from your pockets

ANSWER: c

58. A tornado cloud that does not touch the ground is called:
a. a funnel cloud
b. a wall cloud
c. a roll cloud
d. a mesocyclone

ANSWER: a

59. A funnel cloud is composed primarily of:
a. cloud droplets
b. dust and dirt from the ground
c. raindrops
d. hail
e. ice crystals

ANSWER: a

60. Which of the following statements about tornadoes is correct?
a. all tornadoes rotate in a counterclockwise direction
b. tornadoes never strike the same place twice
c. all tornadoes make a distinction roar
d. the United States has more tornadoes that any other country in the world

ANSWER: d

61. The rotating updraft inside a severe thunderstorm is called a:
a. mesohigh
b. mesocyclone
c. suction vortex
d. funnel cloud
e. roll cloud

ANSWER: b

62.   The small, rapidly rotating whirls that sometimes occur within a large tornado are called:
       a. microtornadoes
       b. whirl winds
       c. suction vortices
       d. mesocyclones

ANSWER: c

63.   A funnel cloud or tornado may develop from this rotating cloud that extends beneath a severe
       thunderstorm.
       a. mammatus cloud
       b. anvil cloud
       c. roll cloud
       d. wall cloud
       e. suction vortex

ANSWER: d

64.   The funnel cloud characteristic of a tornado is <u>principally</u> formed by:
       a. condensation of water vapor in air drawn into the low pressure core of the
          tornado
       b. dust and dirt picked up from the atmosphere
       c. clouds being funneled by downward air currents coming out of a cumulonimbus
          cloud
       d. water drawn up from the sea surface into the cloud
       e. gaseous products from intense lightning activity within the tornado

ANSWER: a

65.   In a typical tornado the winds are usually not much more than ___miles per hour.
       a. 25
       b. 50
       c. 100
       d. 200

ANSWER: c

66.   Most tornadoes move from:
       a. northwest to southwest
       b. southwest to northeast
       c. south to north
       d. southeast to northwest

ANSWER: b

67.   Which of the following factors is most important in determining the strength of a tornado?
        a. diameter
        b. air temperature
        c. duration
        d. central pressure

ANSWER: d

68.   If a tornado is rotating in a counterclockwise direction and moving toward the northeast, the strongest winds will be on its ___ side.
        a. southwestern
        b. southeastern
        c. northeastern
        d. northwestern

ANSWER: b

69.   A typical diameter of a tornado would be:
        a. 50 meters
        b. 250 meters
        c. 1000 meters
        d. 2500 meters
        e. 4000 meters

ANSWER: b

70.   The so-called "Tornado Belt" of the United States is located:
        a. in Florida
        b. in the Central Plains
        c. in the Ohio Valley
        d. in the middle Atlantic states
        e. along the Gulf Coast

ANSWER: b

71.   Different tornadoes spawned by the same thunderstorm are said to occur:
        a. in unison
        b. in sequence
        c. in repetition
        d. in families

ANSWER: d

72.   The most frequent time of day for tornadoes to form is in the:
        a. early morning just after sunrise
        b. late morning just before noon
        c. evening just after sunset
        d. afternoon
        e. middle of the night

ANSWER: d

73.   In the United States, tornadoes are most frequent during the __, and least frequent during the __.
        a. summer, fall
        b. spring, winter
        c. spring, fall
        d. summer, winter
        e. winter, fall

ANSWER: b

74.   Which of the following regions would you expect to have the most tornadoes in the winter?
        a. southern Great Plains
        b. Oklahoma
        c. northern Great Plains
        d. southern Gulf States

ANSWER: d

75.   Tornadoes are usually observed:
        a. behind cold fronts
        b. on the windward side of mountains
        c. near large bodies of water
        d. along occluded fronts
        e. ahead of cold fronts

ANSWER: e

76.   In an eastward moving thunderstorm, the most likely place for a tornado to develop is in the part of the storm.
        a. northeast
        b. southeast
        c. northwest
        d. southwest

ANSWER: d

77. In a region where severe thunderstorms with tornadoes are forming, one would not expect to observe:
    a. a strong ridge of high pressure over the region
    b. a dry tongue of cold air between the 700 and 500 mb levels
    c. the polar jet stream above the region
    d. moist warm air moving north at about the 850 mb level

ANSWER: a

78. Which below is the best indication that a severe thunderstorm is about to produce a tornado?
    a. a wall cloud
    b. a roll cloud
    c. a mammatus cloud
    d. a gust front

ANSWER: a

79. At home, when confronted with an approaching tornado, you should:
    a. open the windows right away
    b. grab a video camera and start filming
    c. listen to see whether the tornado has an audible roar
    d. see shelter immediately

ANSWER: d

80. The Fujita scale pertains to:
    a. the size of a tornado producing thunderstorm
    b. the amount of hail that falls from a mature thunderstorm
    c. the size of the thunderstorm image on a radar screen
    d. the strength of a tornado

ANSWER: d

81. About two-thirds of all tornadoes fall into which of the following categories on the Fujita scale?
    a. F0 or F1
    b. F2 or F3
    c. F0 or F5
    d. F4 or F5

ANSWER: a

82. Damage to structures inflicted by tornadoes can be caused by:
    a. flying debris
    b. the tornadoes high winds
    c. the drop in air pressure as a tornado moves overhead
    d. the drop in air pressure above a roof as high winds blow over it
    e. all of the above

ANSWER: e

83. After a tornado is spotted:
    a. a tornado watch is issued
    b. its direction of movement is carefully monitored by aircraft
    c. a tornado warning is issued
    d. hail begins to fall from the cloud
    e. attempts may be made to change its direction of movement

ANSWER: c

84. A hook-shaped echo on a radar screen often indicates:
    a. a thunderstorm with very frequent lightning
    b. a developing hurricane
    c. the possible presence of a tornado-producing thunderstorm
    d. a rotating anvil cloud at the top of a thunderstorm

ANSWER: c

85. The instrument that measures the speed at which precipitation is moving toward or away from an observer is:
    a. the radiosonde
    b. the aerovane
    c. the wind psychrometer
    d. Doppler radar

ANSWER: d

86. The signal detected by a Doppler radar is:
    a. a radiowave emitted by lightning
    b. a soundwave produced by thunder
    c. a radiowave reflected by precipitation
    d. a soundwave produced by wind shear

ANSWER: c

87. A single Doppler radar is not able to:
    a. measure the speed of falling precipitation
    b. measure the speed at which precipitation is moving horizontally
    c. measure the speed at which precipitation is moving parallel to the radar antenna
    d. detect areas of precipitation
    e. detect a mesocyclone

ANSWER: c

88. A Doppler radar determines precipitation ___ by measuring changes in
    the ___ of the reflected radiowave.
    a. size, intensity
    b. velocity, intensity
    c. velocity, frequency
    d. size, frequency

ANSWER: c

89. On a Doppler radar screen, a tornado vortex signature (TVS) appears as:
    a. a region of low pressure
    b. a region of intense precipitation
    c. a region of rapidly changing wind speeds
    d. a region of intense lightning activity

ANSWER: c

90. Most waterspouts:
    a. form in severe thunderstorms
    b. draw water up into their core
    c. have rotating winds of less than 45 knots
    d. form in an area where winds are descending from a cloud
    e. actually form over land

ANSWER: c

91. Which is not a characteristic of a "fair weather" waterspout?
    a. is usually smaller than an average tornado
    b. tends to move more slowly than a tornado
    c. is usually less intense than a tornado
    d. generally indicate stable atmospheric conditions

ANSWER: d

92. A tornado-like event that forms over water is a:
    a. mesohigh
    b. roll cloud
    c. microburst
    d. waterspout
    e. squall line

ANSWER: d

93. Most waterspouts would fall into which category of the Fujita scale?
    a. F_0
    b. F_1
    c. F_2
    d. F_4 or F_5

ANSWER: a

94 Light flashes darting upward from the tops of thunderstorms are called
    a. bead lightning
    b. ball lightning
    c. heat lightning
    d. sprites and Mountain Dew
    e. sprites and blue jets

ANSWER: e

Essay Exam Questions

1. List and discuss some of the atmospheric conditions that are needed for a thunderstorm to develop.

2. List and describe the stages of development of an ordinary thunderstorm. About how long does a single ordinary thunderstorm cell last?

3. Where does the energy contained in a mature thunderstorm come from?

4. In what ways are severe thunderstorms different from ordinary thunderstorms? What are some of the meteorological or atmospheric conditions that favor the development of severe thunderstorms?

5. How does a thunderstorm gust front form? What might you expect to see and feel if a gust front were to approach and pass you on the ground?

6. What is wind shear? Why does wind shear represent a hazard to aviation?

7. Sketch a mature thunderstorm. On your sketch indicate approximately the altitude of the cloud's base and top and show where you might expect to see the anvil cloud, a pileus cloud, mammatus cloud, and a roll cloud. With arrows, indicate where the updraft and downdraft might be found in the cloud. Where would you expect to find strong vertical wind shear? Indicate where the largest concentrations of positive and negative charge would be found in the cloud.

8. What is a squall line? Where would you expect squall lines to form?

9. Where do thunderstorms form most frequently in the US? Why is this the case? Is this also where most tornadoes occur? Explain.

10. List and describe the sequence of events that occur during a cloud-to-ground lightning discharge.

11. Do invisible tornadoes exist? Explain.

12. Compare some damaging straight-line winds with tornado winds. Which do you think are more destructive?

13. What makes a tornado-producing thunderstorm different from other thunderstorms?

14. The region of greatest tornado activity shifts northward from early spring to summer. Why does this occur?

15. Most tornadoes move from the southwest toward the northeast. Why is this true?

116. The fact that thunder sometimes sounds like a loud crash or peal, and other times sounds like a dull rumble, has nothing to do with the intensity of the lightning strike that produced the thunder. Explain why.

Chapter 16
Hurricanes

Summary

Tropical weather and hurricanes are discussed in this chapter. Several unusually strong and destructive hurricanes have affected the US in the past several years and this subject should be especially relevant and interesting to students.

The chapter begins by mentioning some of the differences between atmospheric conditions and storms in the tropics and at middle latitudes. Horizontal temperature and pressure gradients are weaker in the tropics, for example, than at middle latitudes. Moisture is abundant in the tropics, however, and zones of convergence such as might be found associated with an easterly moving wave or near the ITCZ can cause thunderstorms to form.

The chapter next examines the special conditions that make it possible for a mass of thunderstorms to become organized and develop into or 'trigger' a tropical storm or hurricane. Hurricanes form above and derive much of their energy from a warm surface layer of ocean water. Hurricanes generally form between 5 and 20 degrees latitude, where steering winds will cause an overall east to west movement. Schematic diagrams of the air motions in a mature hurricane and characteristic hurricane features including the eye, eye wall and spiral bands, are given. Tropical storm Allison, which caused much devastation throughout the eastern United States in May-June, 2001, is featured in a focus section.

Some of the unique hazards associated with hurricanes are discussed. In addition to powerful winds and flooding caused by heavy rainfall, extensive damage can be caused by the hurricane storm surge. Several recent examples of destructive and costly hurricanes are given, and we see that hurricane forecasting remains an inexact science. Lists of proposed names for future Northern Atlantic and Eastern Pacific hurricanes are included.

Key Terms

tropical storm
Ekman transport
swells
storm surge
spin-up vortices
hurricane watch
hurricane warning
landfall
Saffir-Simpson scale
non-squall cluster

tropical squall cluster
 (squall line)
streamlines
easterly (tropical) wave
hurricane
typhoon
baguio
cyclone
eye (of hurricane)
rain bands

eye wall
trade wind inversion
organized convection
 theory
conditional instability
 of the second kind
 (CISK)
tropical disturbance
tropical depression

Teaching Suggestions

1.   Video footage of recent hurricanes is available for purchase from a variety of sources. See a recent issue of *Weatherwise* magazine for advertisements.

A one-hour program about hurricanes has also been produced by the NOVA television series. This program includes dramatic footage from recent hurricanes.

Student projects

1.   Depending on the time of the year students may be interested in plotting and following hurricane or tropical storm trajectories on a tracking chart. Students can investigate how large, synoptic-scale weather features affect the storm's movement.

2.   Have students prepare a report on a strong recent hurricane (a suitable choice could be made from the list of retired names in the text) or a hurricane of historical interest. The 20 deadliest hurricanes to have struck the US in the 20th century, for example, are listed by J. Williams (*The Weather Book*, Vintage Books, 1997).

Blue Skies 3.   Using the Hurricane/Hurricane Forecasting section of the BlueSkies cdrom, examine the movement of the archived storm Henry. How might its movement have been forecast?

Blue Skies 4.   Using the Hurricane/Virtual Hurricane section of the BlueSkies cdrom, make separate graphs of wind speed vs. distance from the eye as you approach the eye from the north, south, east and west. Does the relationship of wind speed vs. distance change with your direction of approach? Try to explain your results.

Answers to Questions for Thought

1.   In May the surface ocean water in the subtropics is still relatively cool from the winter. In October the water is still warm from the summer.

2.	It would be possible if the water temperature remained warm enough. During most years the water is too cold to generate hurricanes and so the hurricane season officially ends on November 30. However, during a strong El Niño event, the water may remain warm into December. As an example, Hurricane Winnie formed over the tropical eastern Pacific on December 4, 1983.

3.	The winds would decrease more quickly when the hurricane moves over land. The storm would be deprived of its primary source of energy (latent heat of condensation) and friction with the surface would disrupt its circulation and slow its winds.

4.	The rapid rate of evaporation cools the water.

5.	Veering to the south of the storm. The hurricane's strongest winds would be found in its northern sector.

6.	It would not necessarily begin with the letter C because several tropical storms may have formed but not developed into full-blown hurricanes before this.

7.	To the east. Winds circulate clockwise around tropical cyclones in the Southern Hemisphere, so the onshore winds (and the storm surge) will arrive to the east of Darwin.

8.	The hurricane's dissipation will be somewhat slowed if it encounters a mountain range after making landfall, because orographic lifting will provide some vertical motions to the air. The most damage is likely to be from flooding and flash flooding resulting from heavy rains.

9.	Possible reasons include: the hurricane can pass over an area of warm surface ocean water, it can move away from the equator where the Coriolis force is stronger, it can experience decreasing levels of vertical wind shear, the trade wind inversion can weaken, upper-level divergence can increase.

Answers to Problems and Exercises

1.	(a) The winds should change from E to NE to N to NW.
	(b) Probably the strongest winds would occur with the east or northeast wind coming from the ocean as surface friction would tend to slow down the wind blowing from other directions.
	(c) The lowest sea level pressure would most likely occur with the north wind. At this point the storm would be directly offshore, its point of closest approach.

2.	(a) Category 3
	(b) Category 3
	(c) Category 4
	(d) Category 4

Multiple Choice Exam Questions

1. A weak trough of low pressure found in the tropics, along which hurricanes occasionally form, is called:
    a. an open wave
    b. an easterly wave
    c. a baroclinic wave
    d. a permanent wave

ANSWER: b

2. Streamlines on a weather map depict:
    a. water temperature
    b. pressure
    c. wind flow
    d. dew point
    e. ocean currents

ANSWER: c

3. Which of the following is not true concerning an easterly wave?
    a. moves from east to west
    b. has converging winds on its western side
    c. showers and thunderstorms may be found on its eastern side
    d. indicates a region of lower-than-average pressure

ANSWER: b

4. Suppose the eye of a hurricane passed directly over you, and you survived the experience. If winds were from the northeast as the eyewall first approached you, from what direction did the winds blow when the eyewall reached you the second time?
    a. NW
    b. NE
    c. SE
    d. SW

ANSWER: d

5. Pressure at the center of a hurricane is ___ than the surroundings at the surface and ___ than the surroundings aloft.
    a. higher, lower
    b. lower, higher
    c. lower, lower
    d. higher, higher

ANSWER: b

6.   The vertical structure of the hurricane shows an upper-level ___ of air, and a surface ___ of air.
      a. outflow, inflow
      b. outflow, outflow
      c. inflow, outflow
      d. inflow, inflow

ANSWER: a

7.   In the Northern Hemisphere, hurricanes and middle latitude cyclones are similar in that both:
      a. have surface weather fronts
      b. intensify with increasing height above the ground
      c. have winds that blow counterclockwise around their centers
      d. will generally move from west to east

ANSWER: c

8.   Which would you not expect to observe as the eye of a hurricane passes directly over your area?
      a. an increase in surface temperature
      b. a very low surface pressure reading
      c. high winds
      d. little or no precipitation

ANSWER: c

9.   Most hurricanes have fronts.
      a. true
      b. false

ANSWER: b

10.   In a hurricane, the eye wall represents:
      a. the exact center of the storm
      b. the area of broken cloudiness at the center
      c. a region of light winds and low pressure
      d. a zone of intense thunderstorms around the center
      e. a layer of cirrus cloud in the center of the storm

ANSWER: d

11.   The strongest winds in a hurricane are found:
      a. at the center of the storm
      b. in the eye wall
      c. in the rain bands
      d. at upper levels, above the center of the hurricane
      e. near the periphery of the hurricane

ANSWER: b

12. Hurricane winds rotate in a clockwise direction:
    a. in the Northern Hemisphere only
    b. in the Southern Hemisphere only
    c. in both the Northern and Southern Hemispheres
    d. in neither hemisphere

ANSWER: b

13. Although much is not yet understood about hurricanes, meteorologists have a reasonably complete understanding of *how hurricanes form*.
    a. true
    b. false

ANSWER: b

14. At the periphery of a hurricane the air is ___, and several kilometers above the surface, in the eye, the air is ___.
    a. sinking, sinking
    b. sinking, rising
    c. rising, sinking
    d. rising, rising

ANSWER: a

15. In order for a hurricane to form, the layer of warm water must extend from the ocean surface to a depth of at least 400 meters.
    a. true
    b. false

ANSWER: b

16. As a northward-moving hurricane passes to the east of an area, surface winds should change from:
    a. NW to N to NE
    b. W to SW to S
    c. NE to N to NW
    d. S to SW to W
    e. SE to S to SW

ANSWER: c

17.   The main reason hurricanes don't develop over the south Atlantic Ocean adjacent to South America is because:
        a. the Coriolis force is too small there
        b. the pressure gradient force is too weak in that area
        c. the surface water temperatures are too cold
        d. the air at the surface is always diverging

ANSWER: c

18.   Hurricanes do not form:
        a. along the ITCZ
        b. along the equator
        c. with an easterly wave
        d. when the trade wind inversion is weak
        e. when the surface water temperature exceeds 25 °C

ANSWER: b

19.   As surface air rushes in toward the eye of a hurricane, the air expands and should cool. The main reason the surface air is not cooler around the eye is because:
        a. the sinking air near the eye warms the air
        b. friction with the water adds heat to the air
        c. the warm water heats the air
        d. sunlight heats the air

ANSWER: c

20.   The main source of energy for a hurricane is the:
        a. upper-level jet stream
        b. rising of warm air and sinking of cold air in the vicinity of weather fronts
        c. warm ocean water and release of latent heat of condensation
        d. ocean currents and tides

ANSWER: c

21.   Which below only forms over water?
        a. thunderstorms
        b. funnel clouds
        c. tornadoes
        d. mesocyclones
        e. hurricanes

ANSWER: e

22.   Hurricanes dissipate when:
        a. they move over colder water
        b. they move over land
        c. surface inflow of air exceeds upper-level outflow of air
        d. all of the above

ANSWER: d

23.   Which below is <u>not</u> an atmospheric condition conducive to the formation of hurricanes?
        a. a region of converging surface winds at the surface
        b. warm water
        c. strong upper-level winds
        d. cold air aloft
        e. moist, humid surface air

ANSWER: c

24.   Just before a storm becomes a fully developed hurricane, it is in the _____ stage. a. tropical depression
        b. tropical disturbance
        c. tropical storm
        d. cyclone
        e. typhoon

ANSWER: c

25.   The main difference between a hurricane and a tropical storm is that:
        a. hurricanes are larger
        b. tropical storms are more than 500 miles from the US mainland
        c. winds speeds are greater in a hurricane
        d. hurricanes have a clearly defined eye on satellite photographs

ANSWER: c

26.   The first three stages of a developing hurricane are (from first stage to third stage):
        a. tropical disturbance, tropical storm, typhoon
        b. tropical depression, tropical disturbance, tropical storm
        c. tropical disturbance, tropical depression, tropical storm
        d. cyclone, typhoon, tropical storm

ANSWER: c

296

27.   Which statement below is not correct concerning hurricanes?
        a. they may contain tornadoes
        b. they may contain severe thunderstorms
        c. a hurricane moving northward over the Pacific will normally survive for a
           longer time than one moving north over the Atlantic
        d. a weakening hurricane can move up to middle latitudes and turn into an
           extratropical cyclone

ANSWER: c

28.   Which method below describes how scientists have tried to modify hurricanes?
        a. putting an oil slick over the ocean water and igniting it
        b. seeding the hurricanes with silver iodide
        c. igniting huge smoke bombs in the eye of the storm
        d. seeding the hurricanes with hair-thin pieces of aluminum called chaff.

ANSWER: b

29.   Which of the following areas in the United States would most likely experience thunderstorms,
        hurricanes and tornadoes during the course of one year?
        a. Pacific Coast states
        b. New England states
        c. Gulf Coast states
        d. Great Plains states

ANSWER: c

30.   On the Saffir-Simpson scale a category 5 storm would indicate:
        b. a weak hurricane
        b. a moderately strong hurricane
        c. a very strong hurricane
        d. none of the above, the Saffir-Simpson scale applies to tornadoes

ANSWER: c

31.   Which below is the best indication that a hurricane will likely strike your area within 24 hours?
        a. a hurricane watch issued by the National Weather Service
        b. high cirrus clouds moving in from the east
        c. a hurricane warning issued by the National Weather Service
        d. easterly or northeasterly winds with speeds in excess of 30 knots
        e. a rapid drop in pressure and heavy rains

ANSWER: c

32. On the Saffir-Simpson hurricane scale, a hurricane with winds in excess of 155 mi/hr (135 knots) and a central pressure of 910 mb (26.87 in.) would be classified as a category ___ hurricane.
   a. 1
   b. 2
   c. 3
   d. 4
   e. 5

ANSWER: e

33. Hurricanes that move into India and Australia are usually called ___ in this part of the world.
   a. typhoons
   b. hurricanes
   c. cyclones
   d. extratropical cyclones

ANSWER: c

34. The strongest winds in a hurricane heading westward toward Florida would most likely be found on the ___ side.
   a. northern
   b. southern
   c. eastern
   d. western

ANSWER: a

35. Most of the destruction caused by a hurricane is due to:
   a. high winds
   b. flooding
   c. tornadoes
   d. hail

ANSWER: b

36. A strong trade wind inversion _____ to the development of a hurricane.
   a. is conducive
   b. is not conducive
   c. is immaterial

ANSWER: b

37.   A hurricane warning:
        a. gives the exact location where a hurricane will make landfall
        b. is usually issued several days ahead of a hurricane's arrival
        c. gives the percent chance of a hurricane's center passing within 65 miles of a
           community
        d. is issued when a hurricane approaches to within 500 miles of the US mainland
        e. is issued whenever surface wind speeds exceed 74 mi/hr.

ANSWER: c

38.   The ___ is a measure of hurricane strength based on hurricane winds and central pressure.
        a. Fujita scale
        b. Richter scale
        c. Saffir-Simpson scale
        d. Beaufort scale

ANSWER: c

39.   A storm of tropical origin whose high winds and water cause a great deal of destruction to islands
        in the western North Pacific is (in this part of the world) known as a:
        a. hurricane
        b. cyclone
        c. willy willy
        d. mesocyclone
        e. typhoon

ANSWER: e

40.   An intense storm of tropical origin that forms over the Pacific Ocean adjacent to the west coast of
        Mexico would be called a:
        a. hurricane
        b. typhoon
        c. cyclone
        d. willy willy
        e. baguio

ANSWER: a

41.   Which of the following is true?
        a. hurricanes are only given male names
        b. hurricanes are only given female names
        c. hurricanes are alternately assigned male and female names
        d. Atlantic hurricanes are given male names and Pacific hurricanes are given
           female names

ANSWER: c

42. Suppose that 5 of this year's tropical storms develop into hurricanes over the north Atlantic Ocean. The name of the third hurricane would start with the letter ___.
    a. A
    b. B
    c. C
    d. D
    e. impossible to tell

ANSWER: e

43. Storms that form in the tropics are given names when:
    a. they reach tropical storm strength
    b. they become fully developed hurricanes
    c. they approach to within 250 miles of land
    d. rotation becomes visible on a satellite photograph

ANSWER: a

44. Hurricanes can't form along the equator because
    a. there isn't much water along the equator, it's mostly land
    b. the Coriolis force is too small along the equator
    c. there are no feedback mechanisms along the equator
    d. it's too humid for the ocean water to evaporate

ANSWER: b

Essay Exam Questions

1. In what ways are weather conditions in the tropics and at middle latitudes different?

2. With sketches show the structure of a mature hurricane as it would appear from the side and from above. Indicate and label the major features.

3. Why can't hurricanes form in the midlatitudes?

4. Very heavy rainfall amounts are often recorded when a hurricane or tropical storm moves over land. Why do these storms produce so much rain?

5. List and describe some of the conditions that are favorable to hurricane development. What atmospheric conditions inhibit hurricane formation and growth?

6. Would you expect hurricanes in the Southern Hemisphere to be any different from hurricanes in the Northern Hemisphere?

7. Hurricane season for the tropical North Atlantic and North Pacific oceans normally runs from June through November. Why don't hurricanes form in these locations at other times of the year?

8. Describe a hurricane as completely as you can by giving typical values for the following characteristics: diameter, location and size of the eye, direction of rotation and speed of winds, central pressure, direction and speed of movement, duration. When and where do hurricanes form?

9. Where is the Bermuda high located during the summer and fall? How might the path of a hurricane, moving toward the west from Africa, be affected by the Bermuda High as the hurricane approaches the United States?

10. Compare the size of an average hurricane to that of an average tornado. Given the size and intensity of each, which do you think is the more destructive storm? Why?

11. In your opinion do hurricanes pose the greatest threat to the east or west coast of the United States? Explain.

12. Even though more hurricanes form, on average, over the Eastern Pacific than over the tropical North Atlantic, we generally hear less about them. Why do you think this is so?

13. In recent years, the number of deaths caused by hurricanes has decreased, but the cost of hurricane damage has increased. How would you explain this?

14. Why are hurricanes so destructive? List some of the hazards associated with a hurricane.

15. What is a storm surge? How does a storm surge form?

16. Compare and contrast the main features of a hurricane with those of a strong mid-latitude cyclone.

Chapter 17
Air Pollution

Summary

Air pollutants and air pollution meteorology are examined in this chapter. Many concepts introduced earlier in the text, such as atmospheric stability and small scale wind circulation patterns, are relevant to the air pollution problem and are integrated into the discussion.

The chapter begins with a brief historical review of air pollution. Many students will be surprised, perhaps, to see that concern over air pollution dates from at least the 13th century. Students may also be unaware of the very serious air pollution events that occurred earlier in this century in Europe and the United States. The London smog of 1952, for example, remains today the world's worst air pollution disaster.

Sources and environmental effects of the primary air pollutants, carbon monoxide, sulfur dioxide, nitrogen oxides, particulate matter and volatile organic compounds are discussed next. The characteristics of ozone is presented in some detail. A distinction is made between tropospheric and stratospheric ozone, with the main features, impacts and chemistry details presented for each. A quantitative measure of air quality, the air quality index, is examined. We see that, while air quality has improved following the passage of clean air legislation, many urban areas still frequently exceed these standards.

The roles that weather conditions, local topography, and the urban environment play in air pollution are discussed. The chapter includes a brief discussion of indoor air pollution and concludes with a discussion of acidic deposition.

Key Terms

air pollutants
fixed sources
mobile sources
primary air pollutants
secondary air pollutants
particulate matter
aerosols
PM-10
arctic haze
carbon monoxide (CO)
Environmental Protection
  Agency (EPA)
hemoglobin
sulfur dioxide (SO$_2$)
sulfur trioxide (SO$_3$)
sulfuric acid (H$_2$SO$_4$)
volatile organic
  compounds (VOCs)
hydrocarbons
carcinogens
nitrogen dioxide (NO$_2$)
nitric oxide (NO)
oxides of nitrogen (NO$_x$)
nitric acid (HNO$_3$)
smog
photochemical smog
Los Angeles-type
  smog
London-type smog
ozone (O$_3$)

stratospheric ozone
tropospheric ozone
hydroxyl radical (OH)
PAN (peroxyacetyl
  nitrate)
chlorofluorocarbons
chlorine
chlorine monoxide
Montreal Protocol
bromine
ozone hole
National Ozone
  Expedition (NOZE)
polar vortex
polar stratospheric
  clouds
photochemical oxidants
polycyclic aromatic
  hydrocarbons (PAHs)
radon
radon detectors
polonium
formaldehyde
asbestos
asbestosis
"sick building
  syndrome"
primary ambient air
  quality standards
secondary standards

nonattainment area
air quality
  index (AQI)
radiation (surface)
  inversion
fanning smoke plume
fumigation
looping smoke plume
coning smoke plume
lofting smoke plume
subsidence inversion
mixing layer
mixing depth
atmospheric
  stagnation
urban heat island
country breeze
METROMEX (Metropolitan
  Meteorological
  Experiment)
BASIN (Basic Studies
  on Airflow, Smog and
  Inversion)
dry deposition
wet deposition
acid rain
acid deposition
acid fog

Teaching Suggestions

1. Heat a small piece of coal using a propane torch in class. The burning coal will produce a lot of sooty black smoke and a pungent, sulfurous odor. Students will get a better appreciation of how unhealthy conditions must have been during the London smog episodes.

2. Effective classroom demonstrations of some of the common reactions in air pollution chemistry have been given by J. L. Hollenberg *et al.* (ref: "Demonstrating the Chemistry of Air Pollution," *J. Chem. Educ., 64*, 893-894, 1987).

Student Projects

1.   A daily air quality summary is given in many local papers. Have students report on conditions in their location. What emissions are of most concern? What are the primary sources of these emissions? How do the reported values correlate with observed atmospheric conditions, visibility, and synoptic weather conditions?

2.   Have students investigate and report on measures are being taken locally to reduce air pollution. This might include car pooling, subsidized bus passes or other use of mass transit, use of oxygenated fuels during certain times of the year, vehicle emissions inspections, and no-burn nights.

Blue Skies   3.   Use the Weather Analysis/Find the Front section of the BlueSkies cdrom to locate an area of stagnation that might be subject to an air pollution episode, if the necessary pollution emissions were present.

Blue Skies   4.   Use the Atmospheric Chemistry/Smog section of the BlueSkies cdrom to answer the following question. Is it possible to have low values of ozone (< 20 ppb) with high concentrations (> 4 ppb) of NO_x? Explain.

Answers to Questions for Thought

1.   (a) No. Ozone is produced by photodissociation of nitrogen dioxide (NO_2) and subsequent combination of atomic and molecular oxygen. Photodissociation of NO_2 also produces nitric oxide (NO) which reacts with and destroys ozone.
         (b) Hydrocarbons can react with hydroxyl radicals and then molecular oxygen. The product reacts with NO to form NO_2. Hydrocarbons thus prevent NO from reacting with and destroying ozone.

2.   In the afternoon when the surface air is highest and the atmosphere is most unstable.

3.   Radiation inversions are usually shallow and solar heating causes them to weaken and disappear by mid-day. Subsidence inversions may persist for several days.

4.   Clear skies will allow strong radiational cooling at night and a radiation inversion may form. With plentiful sunshine, photochemical reactions will produce secondary pollutants such as ozone. A deep high pressure system will generally be accompanied by light surface winds and a subsidence inversion which may remain over an area for several days.

5.   In cities, heat is given off by vehicles and factories as well as by heating and cooling units. Surface cooling in cities is reduced by the tall walls of buildings. What sunlight does penetrate clouds and haze in a city produces warming. In rural areas, a substantial portion of the incident sunlight may be used to evaporate water and doesn't produce warming.

6.   Acid snow can build up over the winter and then be released suddenly into a lake when the snow melts in the spring. Acid rain would be added to a lake more gradually.

7.   A subsidence inversion could produce a stable layer at upper levels and confine mixing to an unstable air layer next to the ground. A subsidence inversion will often persist for several days and would

not disappear during the afternoon as you would expect a radiation inversion to do.

8.   We wish to reduce high ozone concentrations at the earth's surface because ozone is a harmful pollutant that irritates the eyes and the respiratory system, and aggravates chronic respiratory diseases. Although stratospheric ozone is the same gas, at stratospheric altitudes it doesn't come into contact with humans, thus its effects are mainly beneficial (it absorbs UV radiation).

9.   Nighttime inversions would prevent mixing, allowing pollution to become more concentrated. Also, in the absence of convection forced by incoming solar radiation, the polluted air found near the ground will not warm and rise to higher altitudes.

10.   The taller smokestack will likely reduce air quality problems in its immediate vicinity. However, the taller stack will create new air quality problems for areas downwind.

11.   People generally wear clothing when walking in the rain, so most of their skin is protected. Also, most people bathe frequently, thus washing the acidic rain off their bodies.

Answers to Problems and Exercises

2.   (a) The air quality would be considered hazardous.
        (b) General health effects would include aggravation of symptoms of respiratory disease, possible death for ill or elderly people, and a decrease in exercise tolerance for healthy individuals. All persons should avoid outdoor physical activities. Elderly people and people with respiratory and/or heart problems should remain indoors.

3.   Mixing depth will increase.

4.   1000 times more acidic. Each unit on the pH scale corresponds to a factor of 10 change in acidity (i.e., H^+ ion concentration).

Multiple Choice Exam Questions

1.   The following toxic gas was an important component in London's smoke fogs.
        a. ozone (O_3)
        b. sulfur dioxide (SO_2)
        c. radon (Rn)
        d. carbon monoxide (CO)

ANSWER: b

2. The smoke in London smogs came primarily from:
   a. exhaust from diesel engines
   b. trash fires
   c. factories in Eastern Europe
   d. coal combustion

ANSWER: d

3. Which of the following is <u>not</u> true of fine particulate matter (particles less than one micrometer in diameter) in the atmosphere ?
   a. particles may remain suspended in the atmosphere for several weeks
   b. particles are not readily removed from the atmosphere by rain and snow
   c. particles are small enough to penetrate into the lungs
   d. particles can cause a significant reduction in visibility

ANSWER: b

4. Collectively, particles of soot, smoke, dust and pollen are called:
   a. hydrocarbons
   b. aerosols
   c. carcinogens
   d. haze

ANSWER: b

5. Which of the following gases will replace oxygen in blood hemoglobin and thereby reduce the transport of oxygen to the brain?
   a. sulfur dioxide (SO_2)
   b. carbon monoxide (CO)
   c. carbon dioxide (CO_2)
   d. methane (CH_4)

ANSWER: b

6. Which of the following statements is <u>not</u> true of carbon monoxide (CO)?
   a. it replaces oxygen in the blood's hemoglobin
   b. it is removed slowly from the atmosphere
   c. it is produced by the incomplete combustion of carbon-containing fuels
   d. roughly half of the CO in the atmosphere is produced by automobiles

ANSWER: b

7.   Volcanoes are an important natural source of:
        a. benzene
        b. ozone
        c. sulfur dioxide
        d. carbon monoxide
        e. chlorofluorocarbons

ANSWER: c

8.   Which of the following gases is an example of a volatile organic compound or hydrocarbon?
        a. sulfur dioxide
        b. carbon dioxide
        c. methane
        d. ozone

ANSWER: c

9.   Nitrogen dioxide (NO_2) reacts with ___ in the atmosphere to form nitric acid (HNO_3).
        a. hydrogen
        b. water vapor
        c. ozone
        d. carbon dioxide
        e. sulfur dioxide

ANSWER: b

10.   A primary component of photochemical smog is:
        a. ozone
        b. carbon monoxide
        c. sulfur dioxide
        d. chlorofluorocarbons

ANSWER: a

11.   Photochemical smog is also termed:
        a. London-type smog
        b. subsidence smog
        c. mixing layer smog
        d. sea breeze smog
        e. Los Angeles-type smog

ANSWER: e

12. An air quality index value of 35 on a particular day would indicate ___ conditions.
    a. good
    b. unhealthful
    c. extremely hazardous
    d. moderately hazardous

ANSWER: a

13. Overall, the air in Los Angeles is ___ it was 20 years ago.
    a. less polluted today than
    b. more humid today than
    c. about as polluted today as
    d. significantly denser today than
    e. more polluted today than

ANSWER: a

14. The mixing depth is:
    a. a very stable layer that extends from the surface up to the base of an inversion
    b. an unstable layer that extends from the surface up to the base of an inversion
    c. a very stable layer found upward from an inversion
    d. an unstable layer that extends upward from an inversion

ANSWER: b

15. A *looping* shape exhibited by a smokestack plume is indicative of _____ conditions.
    a. inversion
    b. stable
    c. neutral
    d. unstable
    e. both a and b

ANSWER: d

16. A *coning* shape exhibited by a smokestack plume is indicative of _____ conditions.
    a. inversion
    b. stable
    c. neutral
    d. unstable
    e. both a and b

ANSWER: c

308

17. A *fanning* shape exhibited by a smokestack plume is indicative of _____ conditions.
    a. inversion
    b. stable
    c. neutral
    d. unstable
    e. both a and b

ANSWER: e

18. A mixing layer is characterized by:
    a. enhanced vertical air motions
    b. suppressed vertical air motions
    c. strong horizontal winds
    d. high concentrations of pollutants

ANSWER: a

19. Which of the following conditions would act to prevent a buildup of pollutants near the surface?
    a. light surface winds
    b. a strong subsidence inversion
    c. a large, slow-moving anticyclone
    d. a deep mixing layer

ANSWER: d

20. Pollution is most severe in urban areas when:
    a. a cold upper-level low moves into a region
    b. a warm front passes through the area
    c. a large slow-moving anticyclone moves into an area
    d. a storm system begins developing to the west.
    e. a cold front passes through the area

ANSWER: c

21. Atmospheric stagnation is a condition normally brought on by:
    a. thunderstorms
    b. slow-moving anticyclones
    c. overcast skies
    d. tall buildings in a city
    e. movement of an upper level trough overhead

ANSWER: b

22.   Which of the following would probably result in polluted conditions over a fairly long duration?
        a. a radiation inversion
        b. a subsidence inversion
        c. persistent winds
        d. overcast skies

ANSWER: b

23.   The urban heat island is:
        a. warmer air temperatures in urban areas compared to surrounding rural areas
        b. a concentration of energy use in an urban area
        c. locating factories in a single location downwind from cities
        d. use of conservation techniques to reduce energy use in cities

ANSWER: a

24.   Sunshine in a city is typically less intense than in surrounding rural areas because:
        a. of the higher pollution levels in cities
        b. of the air temperature in cities
        c. tall structures in cities block our view of the horizon
        d. cement and asphalt in cities absorb a larger percentage of the incident sunshine

ANSWER: a

25.   When contrasted to a rural area, cities usually do not have higher:
        a. concentrations of pollution
        b. frequency of fog
        c. temperatures
        d. visibilities
        e. frequency of thunderstorms

ANSWER: d

26.   Studies suggest that downwind of a large industrial area:
        a. average annual precipitation can increase
        b. higher surface temperatures are observed
        c. breezy conditions are frequently observed
        d. all of the above

ANSWER: a

27. Which of the following contribute(s) to the formation of an urban heat island?
    a. a large part of the incident sunlight in rural areas is used to evaporate water in vegetation and the soil
    b. heat is released by vehicles in urban areas
    c. heat is released slowly in cities at night
    d. all of the above

ANSWER: d

28. On clear, cold winter nights, cities tend to cool ___ than rural areas and have ___ temperatures.
    a. more slowly, higher
    b. more quickly, higher
    c. more slowly, lower
    d. more quickly, lower

ANSWER: a

29. A country breeze would probably be associated with:
    a. a large high-pressure area that forms over a city
    b. a hot and humid summer day in a large city
    c. a period of heavy rain that falls over a city
    d. a strong urban heat island

ANSWER: d

30. A country breeze blows:
    a. from the city toward the country at night
    b. from the city toward the country during the day
    c. from the country toward the city at night
    d. from the country toward the city during the day

ANSWER: c

31. Which is not correct about acid deposition?
    a. it is only a problem in New England and Scandinavia
    b. it can damage plants and water resources
    c. it is caused mainly by the release of oxides of sulfur and nitrogen
    d. can fall to ground in dry or wet forms

ANSWER: a

32.   Another name for acid rain:
        a. wet deposition
        b. acid fog
        c. dry deposition
        d. "sour" rain

ANSWER: a

33.   The problem of acid rain is probably most severe in which of the following regions:
        a. Gulf Coast
        b. New England
        c. Desert Southwest
        d. Pacific Northwest
        e. Central Plains

ANSWER: b

34.   A decline in the health of forests in Germany has been attributed to:
        a. erosion caused by excessive lumber cutting
        b. acid rain
        c. increased CO_2 concentrations and global warming
        d. urbanization

ANSWER: b

35.   Rain with a pH of 5.6 would be considered:
        a. acidic
        b. alkaline
        c. neutral
        d. polluted
        e. harmful

ANSWER: a

36.   Erosion of limestone buildings, fountains, and sculptures is being caused largely by:
        a. acid rain
        b. ozone
        c. vibrations caused by automobile traffic
        d. urban heat island

ANSWER: a

37.   A greenhouse gas used as a refrigerant, a solvent, and during the manufacture of foam:
       a. water vapor (H_2O)
       b. carbon dioxide (CO_2)
       c. methane (CH_4)
       d. polyvinylchloride (PVC)
       e. chlorofluorocarbons (CFCs)

ANSWER: e

38.   The ozone hole is found in this atmospheric layer:
       a. thermosphere
       b. troposphere
       c. stratosphere
       d. ionosphere
       e. mesosphere

ANSWER: c

39.   About 97% of all ozone in the atmosphere is found in the:
       a. stratosphere
       b. troposphere
       c. mesosphere
       d. exosphere
       e. thermosphere

ANSWER: a

40.   The so-called "ozone hole" is observed above:
       a. the continent of North America
       b. the equator
       c. the continent of Australia
       d. the continent of Antarctica
       e. the continent of Asia

ANSWER: d

41.   The term "ozone hole" refers to a ___ decrease in ozone concentration.
       a. permanent
       b. seasonal
       c. monthly
       d. daily

ANSWER: b

42. If the concentration of ozone were to decrease significantly, which of the following might also occur?
    a. less absorption of ultraviolet radiation in the stratosphere.
    b. an increase in the number of cases of skin cancer
    c. the stratosphere would cool
    d. more ultraviolet radiation would be absorbed at the earth's surface
    e. all of the above

ANSWER: e

43. Polar stratospheric clouds are thought to contribute to the problem of:
    a. global warming
    b. ozone destruction
    c. acid rain
    d. photochemical smog
    e. radio wave interference

ANSWER: b

44. When chlorofluorocarbons are subjected to ultraviolet radiation, ____ is released which rapidly destroys ozone.
    a. chlorine
    b. nitrogen
    c. carbon dioxide
    d. carbon
    e. water vapor

ANSWER: a

45. What gas is produced naturally in the stratosphere and is also a primary component of photochemical smog in polluted air at the surface?
    a. carbon dioxide
    b. carbon monoxide
    c. ozone
    d. nitrogen dioxide
    e. hydrocarbons

ANSWER: c

46. Which of the following is a major way in which chlorofluorocarbons can enter the stratosphere?
    a. from the exhaust of high-altitude aircraft
    b. in an inversion
    c. in building thunderstorms that penetrate into the lower stratosphere
    d. from the rupture of radiosonde balloons

ANSWER: c

47.   Once released into the atmosphere, chlorofluorocarbons remain about:
       a. 10 days
       b. 100 years
       c. 1 year
       d. 1 month

ANSWER: b

48.   Particulate pollution with diameters less than 2.5 micrometers are particularly dangerous because
       a. they can penetrate deep into the lungs
       b. they dissolve easily in water
       c. they are chemically neutralized by dissolved carbon dioxide
       d. both b and c
       e. none of the above

ANSWER: a

49.   Which of the following are capable of destroying ozone in the stratosphere?
       a. oxygen atoms
       b. chlorine atoms
       c. other ozone molecules
       d. all of the above

ANSWER: d

50.   Thunderstorms are most likely to form on a day when smokestack plumes have a
       a. fanning shape
       b. looping shape
       c. coning shape
       d. lofting shape

ANSWER: b

Essay Exam Questions

1.   List several atmospheric pollutants. What are the most important sources and environmental or health effects of these pollutants?

2.   What atmospheric conditions would you expect to find associated with a major air pollution episode?

3.   What stability condition is most favorable for improving urban air quality? What time of day or night does this stability condition typically occur?

4.   Explain why locating industrial facilities on the perimeter of a city might not always prevent air pollution in the city.

5. What is the urban heat island effect? How could an urban heat island affect atmospheric conditions nearby?

6. How can topography contribute to pollution in a city or region?

7. Would you expect to find the same environmental conditions upwind and downwind of a large city? Explain.

8. What causes acid rain? What are some of the environmental concerns associated with acid rain? Are problems with acid rain confined to a small area or do they extend over a large region?

9. Describe some of the measures that are being taken to reduce the problem of acid precipitation.

10. Is the ozone that is found near the earth's surface a primary or a secondary pollutant? When during the day would you expect to find the peak concentrations of ozone? What are some of the environmental concerns associated with ozone?

11. How does Los Angeles-type smog differ from London-type smog?

12. List some of the ways that pollutants are removed from the atmosphere. Does this occur quickly or slowly?

13. Describe some of the ways an industrial facility that is constructed in or near a large urban area could attempt to reduce emissions of air pollutants.

14. Describe how the use of a tall smoke stack might improve air quality in the vicinity of a large industrial facility.

15. Would you expect the concentrations of air pollutants to vary significantly during the day in the city where you live? If so, when would you expect to find the highest concentrations of pollutants?

16. Explain how ozone might be thought to have both a beneficial and a detrimental role in the earth's atmosphere.

17. Why doesn't an acidified lake look polluted?

18. Would highly polluted cities tend to have more clouds than less-polluted cities? Explain your answer.

Chapter 18
Global Climate

Summary

This chapter examines the variety of different climatic regions found on the earth. The primary emphasis is on global-scale climates, but micro- and macroscale climates are also mentioned briefly. The chapter recalls, first, the different factors that affect and determine the climate of a particular region. These climatic controls include seasonal and latitudinal variations in incident sunlight, proximity to land or sea, ocean and wind currents, and topographical effects. Global distributions of mean temperature and annual precipitation amounts are presented and form the basis for climate classification using the Köppen system.

Each of the five major climatic types in the Köppen classification system are discussed in detail. Examples of yearly temperature variations and monthly precipitation amounts are given for a representative location in each group and for the major sub-categories. We see that a tropical climate, for example, is characterized by abundant rainfall and very little seasonal variation in mean temperature. Seasonal changes are much larger at middle latitudes. Several different climate zones can be identified at middle latitudes depending on whether summers are warm or cool, dry or moist, and by the severity of the winter. Arid zones are found on the earth in areas dominated by subtropical high pressure systems or in the rain shadow of large mountain ranges. Illustrations of the types of vegetation likely to be found in many of the zones are also given.

Key Terms

microclimate
mesoclimate
macroclimate
global climate
climatic controls
orographic uplift
rain shadow
torrid zone
polar (frigid) zone
temperate zone
Köppen classification
  system
Thornthwaite's system
P/E ratio
P/E index
potential evapo-
  transpiration (PE)
tropical rain forest

tropical wet
  climate (Af)
tropical monsoon
  climate (Am)
tropical wet-and-dry
  climate (Aw)
savanna grass
monsoon
arid climate (BW)
xerophytes
semi-arid climate (BS)
steppe
humid subtropical
  climate (Cfa)
marine climate (Cfb)
dry-summer subtropical
  (Mediterranean) (Cs)
subpolar climate (Dfc)

humid continental
  climate (Dfa, Dfb)
humid continental with
  hot summers (Dfa)
humid continental
  with long cool
  summers (Dfb)
taiga
boreal climate
taiga climate
polar tundra
  climate (ET)
permafrost
tundra vegetation
polar ice cap
climate (EF)
highland climate (H)

Teaching Suggestions

1.   Initially, the variety of labels used for the major groups and subcategories in the Köppen classification system may be confusing to students. Keep a map, such as Figure 18.5, displayed throughout the discussion and use a specific city or region to illustrate each climatic zone.

        Encourage students to learn the classification groups by understanding the differences between them and the cause of those differences. Show how climate depends on latitude by observing the changes that occur as one moves in a line from the equator toward the North Pole at constant longitude. Then examine how climate is modified as one moves from west to east across the United States at a single middle latitude. Show where features in the global circulation such as the subtropical highs and the ITCZ are located at various times during the year and explain how these affect climate.

2.   Present and discuss representative examples (plots of average temperature and monthly precipitation totals) of each of the important climate types. List one or two of the key characteristics that can be used to distinguish between the different climate classifications. Then present some new data, but do not reveal the location where the data was obtained. Ask the students how these data would be classified. Then ask the students where, within a particular region such as the United States, these data might have been obtained. After a period of discussion, reveal the actual location. This is also a good point to be sure that students are familiar with United States and world geography.

Student Projects

1.   Have students collect and prepare a plot of yearly average temperature data and average monthly precipitation totals for their city. How would their town be classified using the Köppen system?

2.   Have students attempt to locate different macroscale climate regions in the city where they live. The students will have to devise a system to be used to identify and classify different climate zones. Students could, for example, compare average conditions in high- and low-lying areas in their town, average conditions near and far from a body of water, or conditions inside a city with conditions in a rural area nearby. Are there significant differences in macroscale climates in their city? Are any differences reflected in the vegetation found at different locations within the city.

3.   Have the students look for different microscale climatic environments. Students could, for example, compare average conditions on the north- and south-facing sides of a large building. Students would have to determine what measurements or observations could be made to identify and classify different microscale environments.

4.   In some locations, students could identify and classify the different climatic zones found at different altitudes on a nearby mountain range. Students could, for example, determine the predominant vegetation types at different levels and attempt to relate this seasonal temperature and precipitation variations.

 5.   Use the Atmospheric Chemistry/Temperature Trends section of the BlueSkies cdrom to examine estimates of future global temperatures. Using the slider bar, set the animation to 120 years into the future. Based on the climate model estimates of future conditions, do you think the global pattern of climate will change substantially? In what way(s)?

 6.   Choose three cities in three different continents, each in a different Köppen climate classification. Using the Weather Forecasting/Forecasting section of the BlueSkies cdrom, record the temperature in each of these cities for five consecutive days. Do the temperatures conform to your expectations? Why or why not?

Answers to Questions for Thought

1.   Cities located east of the Rockies receive moisture from the Gulf of Mexico. The Rockies effectively block Gulf moisture from reaching cities located to the east of the Sierra Nevada Mountains. In addition, the Sierras shield the region east of them from Pacific moisture.

3.   The prevailing westerly winds at this latitude give Boston a continental-type climate.

4.   In polar regions, the ground is frozen except in summer when the upper part thaws. Hence, the summer landscape often turns swampy. During the winter, warmth from a heated building could melt the frozen ground beneath it. This could cause the structure to settle into the ground unevenly. To prevent this, some structures are built upon pilings.

5.   The ground in Cfa climate areas is often drier than the ground in areas with Af climates. Dry ground warms up faster than wet ground, allowing surface temperatures to increase.

7.   In arid desert climates. Here the temperatures are warm enough to support liquid precipitation, but relative humidities are low enough to evaporate the rain before it reaches the surface.

8.   San Francisco is closer to the Pacific Ocean and experiences a maritime climate. Sacramento is inland, and air approaching Sacramento from the west must first cross a mountain range where it may contain less moisture due to orographic lifting and precipitation.

Answers to Problems and Exercises

1.   The climate of the city in Table 18.4 would be classified as Cfa. The city would be located in the southeastern part of the United States. Typical vegetation would include pine forests with some scattered oak trees. The climate of the city in Table 18.5 would be classified as Dfb. This city might be located in the northern states from New England to the Central Plains, or in Southern Canada. In the wetter regions of this climate, one would expect to see native vegetation that includes spruce, pine, fir, and oak.

Multiple Choice Exam Questions

1.   Which of the following refers to day-to-day weather variability?
        a. microclimate
        b. mesoclimate
        c. macroclimate
        d. global climate
        e. none of the above

ANSWER: e

2.   The climate of an area about the size of a town would be described as:
        a. mesoclimate
        b. macroclimate
        c. microclimate
        d. urban climate

ANSWER: a

3.   Which of the following is considered a climatic control?
        a. ocean currents
        b. intensity of sunshine and its variation with latitude
        c. prevailing winds
        d. altitude
        e. all of the above

ANSWER: e

4. Which section of the United States is most likely to experience a persistent subsidence inversion during the summer?
a. Pacific Coastal regions
b. Central Plains
c. Southern Florida
d. Atlantic Coastal regions
e. Midwest

ANSWER: a

5. Which of the following cities would most likely have a dry summer?
a. Chicago, Illinois
b. Denver, Colorado
c. Los Angeles, California
d. Baltimore, Maryland
e. Seattle, Washington

ANSWER: c

6. A rainshadow desert is normally found:
a. in the center of a large surface anticyclone
b. on the back (western) side of a large thunderstorm
c. in polar regions where the air is cold and dry
d. in the center of the ITCZ
e. on the downwind side of a mountain range

ANSWER: e

7. The greatest likelihood of experiencing a dry month of February would be along:
a. the equator
b. the west coast of Mexico (latitude 20 °N)
c. the Southern California coast (latitude 35 °N)
d. the Gulf Coast near New Orleans (latitude 30 °N)
e. the coast of southern Alaska (latitude 60 °N)

ANSWER: b

8. The rainiest places in the world are usually located:
a. downwind from mountain ranges
b. in the region of the subtropical highs
c. on the windward side of mountains
d. in the middle of continents
e. along the western side of continents

ANSWER: c

9.   Which of the following *is not* considered a climatic control?
     a. altitude
     b. mountain barriers
     c. ocean currents
     d. insolation
     e. none of the above

ANSWER: e

10. What do Dallol, Ethiopia, Death Valley, California and El Azizia, Libya have in common?
     a. they are all in third-world countries
     b. they are three of the hottest places on the earth
     c. they are three of the coldest places on the earth
     d. they all have frequent snowstorms

ANSWER: b

11. The earth's rainforests are found in:
     a. humid subtropical (Cfa) climates
     b. tropical wet (Af) climates
     c. tropical wet and dry (Aw) climates
     d. all of the above

ANSWER: b

12. The lowest average temperatures in the world occur in:
     a. North America
     b. Northwestern Europe
     c. the Arctic
     d. the Antarctic
     e. Northern Siberia

ANSWER: d

13. Which of the following help explain why the lowest average temperatures in the world are found in Antarctica?
     a. dry air
     b. high altitude
     c. surface reflects incident sunlight
     d. all of the above

ANSWER: d

14.   The Köppen scheme for classifying climates employs annual and monthly averages of:
        a. temperature and precipitation
        b. precipitation and stream runoff
        c. ocean levels and surface pressure
        d. population density and agricultural output
        e. sunshine and soil type

ANSWER: a

15.   The coldest regions of large continents are usually on the west coast.
        a. true
        b. false

ANSWER: b

16.   In Thornthwaite's P/E index, the "P" and "E" stand for:
        a. precipitation and evaporation
        b. precipitation and energy (available)
        c. potential sunlight and energy
        d. potential sunlight and evaporation

ANSWER: a

17.   According to Köppen's classification of climates, tropical climates are designated by the letter:
        a. A
        b. B
        c. T
        d. D
        e. C

ANSWER: a

18.   In Köppen's system of classifying climates, dry climates are designated by the letter:
        a. A
        b. B
        c. C
        d. D

ANSWER: b

19.   Savanna grass would most likely be associated with which climate type:
        a. tropical rainforest (Af)
        b. tropical monsoon (Am)
        c. semi-arid (BS)
        d. tropical wet-and-dry (Aw)

ANSWER: d

20. The term monsoon refers to:
 a. the short summer rainy season observed in some locations
 b. a short dry season
 c. a period of heavy rainfall produced by thunderstorms
 d. a seasonal shift in wind circulation

ANSWER: d

21. The climate classification for a region with average monthly temperatures that **remain above** 18 °C (64 °F) throughout the year and abundant rainfall, except for a short 1 or 2 **month dry** period, would probably be:
 a. tropical wet (Af)
 b. moist subtropical (Cs)
 c. tropical monsoon (Am)
 d. humid continental (Dfa)

ANSWER: c

22. Which below is <u>not</u> characteristic of a tropical wet climate (Af)?
 a. greater temperature variation between day and night than between the warmest and coolest months of the year
 b. extremely high afternoon temperatures, usually much higher than those experienced in middle latitudes
 c. abundant rainfall all year long
 d. afternoon showers and high humidity

ANSWER: b

23. The tropical wet-and-dry climate is influenced mainly by the:
 a. polar front and the subtropical highs
 b. polar front and the ITCZ
 c. subtropical highs and polar front
 d. ITCZ and subtropical highs
 e. subtropical highs and polar lows

ANSWER: d

24. In a tropical wet-and-dry climate, the dry season occurs with the
 a. high sun period (summer)
 b. low sun period (winter)
 c. close proximity of the ITCZ
 d. all of the above

ANSWER: b

25.   In a dry climate:
        a. there are no plants
        b. potential evaporation and transpiration exceed precipitation
        c. there is no precipitation
        d. air temperatures seldom drop below freezing and days are always hot
        e. winters are short, but cold

ANSWER: b

26.   The most abundant climate type over the face of the earth is:
        a. moist tropical climates
        b. dry climates
        c. moist subtropical mid-latitude climates
        d. moist continental climates
        e. polar climates

ANSWER: b

27.   The highest temperatures in the world occur in:
        a. tropical moist climates
        b. subtropical moist climates
        c. continental moist climates
        d. continental dry climates
        e. subtropical deserts

ANSWER: e

28.   Which climate type would normally have the highest afternoon temperatures during the summer?
        a. A
        b. B
        c. C
        d. D
        e. E

ANSWER: b

29.   One would most likely see xerophytes in which climatic type?
        a. A
        b. B
        c. C
        d. D
        e. E

ANSWER: b

30. The global distribution of precipitation is closely associated with
   a. latitude
   b. distribution of mountain ranges and high plateaus
   c. the general circulation of the atmosphere
   d. all of the above

ANSWER: d

31. Which of the following climatic regions would probably have the <u>largest daily</u> temperature range?
   a. tropical wet climate
   b. arid climate
   c. east coast marine climate
   d. Mediterranean climate

ANSWER: b

32. Deserts that experience low clouds and drizzle tend to be found mainly:
   a. in the center of continents
   b. in the rain shadow of a mountain
   c. on the eastern side of continents
   d. on the western side of continents
   e. on small islands near the equator

ANSWER: d

33. One would most likely experience steppe vegetation in a:
   a. semi-arid climate
   b. humid subtropical climate
   c. marine climate
   d. subpolar climate
   e. tropical wet-and-dry climate

ANSWER: a

34. Locations at middle latitudes with monthly average temperatures in the winter that are below 18 °C (64 °F), but above -3 °C (27 °F) have ___ climates.
   a. humid subtropical
   b. moist continental
   c. semi arid
   d. steppe

ANSWER: a

35. A humid subtropical climate is found in what region of the United States?
    a. Southwest
    b. Southeast
    c. Central Plains
    d. Southern California

ANSWER: b

36. A city located on the east coast of the United States with hot, muggy summers and mild winters (the average temperature of the coldest month is above -3 $^{\circ}$C (27 $^{\circ}$F)) has a ___ climate.
    a. moist subtropical (Cfa) climate
    b. mediterranean (Cs) climate
    c. tropical wet and dry climate (Aw)
    d. moist continental (Dfa)

ANSWER: a

37. To be considered a moist continental climate:
    a. winter snowfall must exceed summer rainfall
    b. monthly average temperatures must fall below -3 $^{\circ}$C (27 $^{\circ}$F) during the winter
    c. total rainfall must exceed 60 inches
    d. a region must be located more than 500 miles from the nearest ocean

ANSWER: b

38. A city on the west coast of the United States with mild winters and a long cool summer would have a ___ climate.
    a. moist subtropical (Cfb)
    b. tropical wet and dry (Aw)
    c. moist continental (Dfa)
    d. semi arid (BS)

ANSWER: a

39. According to Köppen's classification of climates, marine climates are observed in the United States mainly along:
    a. the Pacific Northwest coast
    b. coastal areas of the middle Atlantic states
    c. the coast of New England
    d. the Gulf Coast
    e. coastal margins of Southern California

ANSWER: a

40. Mediterranean climates are characterized by:
    a. cool, wet winters and mild to hot, dry summers
    b. cold, wet winters and hot, humid summers
    c. cold, relatively dry winters and mild, humid summers
    d. mild, wet winters and mild, humid summers
    e. cold, snowy winters and hot, dry summers

ANSWER: a

41. The primary reason for the dry summer subtropical climate in North America is:
    a. that storms do not form in summer
    b. that the air is too cool to produce adequate precipitation
    c. the Pacific high moves north in summer
    d. the ITCZ moves north in summer
    e. the Bermuda high intensifies and shifts northward in summer

ANSWER: c

42. One would not expect to experience a D-type climate in:
    a. Alaska
    b. New England
    c. Southern Canada
    d. South America

ANSWER: d

43. The annual movement of the semipermanent highs and lows has a lot to do with the global distribution of precipitation.
    a. true
    b. false

ANSWER: a

44. Which climate type normally has the largest annual range in temperature?
    a. tropical climates
    b. humid-subtropical climates
    c. polar ice cap climates
    d. subpolar climates
    e. dry climates

ANSWER: d

45.   In an east-west oriented mountain range, a rain shadow could be found on the _____ of the range.
       a. east or west
       b. north or south
       c. both a and b

ANSWER: b

46.   A region with long, very cold winters and average temperatures that exceed 10 °C (50 °F) only one or two months during the summer would have ___ climate.
       a. a subpolar (Dfc)
       b. a polar tundra (ET)
       c. a polar ice cap (EF)
       d. an alpine (An)

ANSWER: a

47.   Thawing of an upper layer of permafrost would most likely be observed in which climatic type?
       a. polar ice cap
       b. humid continental with warm summers
       c. humid subtropical
       d. polar tundra

ANSWER: d

48.   Average monthly temperatures that exceed 0 °C (32 °F) but remain below 10 °C (50 °F) is characteristic of:
       a. the polar ice cap (EF) climate zone
       b. the subpolar (Dfc) climate zone
       c. the polar tundra (ET) climate zone
       d. alpine (An) climate zone

ANSWER: c

49.   Within the continental United States (excluding Alaska), one could observe:
       a. taiga
       b. polar tundra
       c. polar icecap
       d. all of the above

ANSWER: d

50.   Tree growth is not possible in a region where:
        a. annual precipitation is 20 inches or less
        b. average monthly temperatures never exceed 10 °C (50 °F)
        c. average monthly temperatures are sometimes less than 0 °C (32 °F)
        d. evaporation exceeds total annual precipitation

ANSWER: b

51.   When the average temperature of the warmest month averages 0 °C (32 °F) or below, Köppen
        classified this as a ___ climate.
        a. subpolar
        b. polar ice cap
        c. polar tundra
        d. arctic

ANSWER: b

52.   When the average temperature of the warmest month of a region averages above freezing, but
        below 10 °C (50 °F), you might expect to observe what type of vegetation there:
        a. savanna grass
        b. tundra
        c. taiga
        d. none - it's too cold for anything to grow

ANSWER: b

53.   The climate of a city situated high in the mountains of the middle latitudes would, according to
        Köppen, be classified as:
        a. tundra
        b. subpolar
        c. alpine
        d. highland

ANSWER: d

54.   The variety of climatic zones that one observes on a mountain are due primarily to differences in:
        a. atmospheric pressure
        b. relative humidity
        c. rainfall amounts
        d. temperature

ANSWER: d

55. A location in the center of a large, mid-latitude continent is likely to experience
    a. a precipitation maximum during winter
    b. a precipitation maximum during summer
    c. abundant precipitation all year long
    d. little precipitation at any time of year

ANSWER: b

Essay Exam Questions

56. In general, would you expect to find more rainfall on the western side or the Eastern side of the subtropical high pressure centers? How does this affect climatic conditions along the east and west coast of the United States?

57. The hottest places on earth are not found near the equator in tropical wet climates, but, rather, in arid climate regions. Why is this true?

58. Discuss some of the factors that determine where the wettest places in the world are found.

59. What meteorological information is the primary basis for climate classification using the Köppen system?

60. The greatest one-month rainfall total (366 in.) occurred in Cherrapunji, India, in July of 1861. The greatest 24 hour rainfall total (43 in.) in the United States occurred in Alvin, Texas, on July 25, 1979. The greatest 42 minute rainfall total (12 in.) occurred in Holt, Missouri, on June 22, 1947. What meteorological conditions do you think produced these record precipitation amounts?

61. Make a rough sketch of North America and indicate approximately where the following climatic types would be found:
    tropical moist climates   moist subtropical climates
    dry climates   moist continental climates
    polar climates   highland climates
    List the principal characteristics of each climate type and describe the climatic controls that influence the location of each climatic region shown on your map.

62. Describe the climate of and the microclimatic variation within the neighborhood in which you live.

63. What differences exist between tropical wet (Af), tropical monsoon (Am), and tropical wet and dry (Aw) climates? Would you expect to observe any significant differences in types of vegetation in these different climates groups?

64. A yearly rainfall of 14 inches in a hot climate will support only sparse vegetation. The same total amount of rain in north central Canada would support a conifer forest. Why is this possible?

65. How is it possible for regions with a desert climate to be found next to an ocean, which you might expect to be an abundant source of moist air?

66. Where is the Mojave desert located? What factors account for the Mojave desert being found there?

67. Identify and describe the different climatic zones you would observe if you traveled from west to east across the middle of the United States?

68. Compare the climate of your town with that of another location at the same latitude, but in a different continent. Explain any similarities or differences.

69. Seattle, Washington (lat. 47.5 °N) has a coastal mediterranean (Csb) climate. New York, New York (41 °N) has a humid subtropical (Cfa) climate. What differences in climate would you expect to find between these two locations?

70. Describe the differences in climate that would be experienced by cities located upwind and downwind of a large, north-south oriented mountain range.

Chapter 19
Climate Change

Summary

This chapter explores the subject of climate change, and begins with a discussion of some of the experimental techniques, including dendrochronology, analysis of O^{18} and O^{16} ratios in ice and marine sediments, and study of geologic formations, that have been used to infer past climatic conditions. A short history of climate on the earth reveals that large changes in the earth's climate have occurred in the past. Students will be surprised, perhaps, to find that the earth was appreciably warmer 65 million years ago; so warm, in fact, that the polar ice caps did not exist. Ice sheets began to reform about 2 million years ago, however, at the start of the Pleistocene epoch, and advanced as far south as New York as recently as 18,000 to 22,000 years ago. Important changes in climate have occurred during the past 1000 years; the Little Ice Age, for example, between about 1550 and 1850, had important effects on agriculture and living conditions in Europe. Considerable attention is given to global temperature changes that have occurred during the past century.

Several of the suggested causes of climate change are examined. Plate tectonics, for example, might explain climate changes that occur on a time scale of millions of years. Shorter term, 10,000 to 100,000 year, variations may be associated with changes in the earth's orbit around the sun. There is evidence, too, that the sun's output may vary with time and that emissions from volcanoes can have relatively short term, but significant effects on climate.

The present concern that anthropogenic emissions of carbon dioxide and other greenhouse gases may be causing global warming is carefully examined. While there seems to have been a small increase in global average temperatures during the past century, we see that, because the response of the oceans to warmer temperatures and increasing CO_2 amounts and the effects of clouds are not well understood, there are large uncertainties in the computer model predictions of future changes. The thermohaline (or conveyor belt) circulation, for example, is a particularly important component in the global climate system, yet the effects of increasing greenhouse gases on this circulation are not known. Finally, projections of future climate changes, as well as some of the effects that an increase or decrease in average surface temperatures might have are discussed.

Key Terms

alpine glaciers
continental glaciers
CLIMAP (Climate: long-
  range investigation,
  mapping and prediction)
oxygen-isotope ratio
otoliths
dendrochronology
Pleistocene epoch
Ice Age
interglacial period
Eemian interglacial
  periods
Bering land bridge
Younger-Dryas (event)
Holocene epoch
mid-Holocene maximum
climatic optimum

Medieval Climatic
  Optimum
Little Ice Age
"the year without a
  summer"
water vapor-temperature
  rise feedback
positive feedback
  mechanism
runaway greenhouse effect
snow-albedo feedback
negative feedback
  mechanism
runaway ice age
theory of plate tectonics
theory of continental
  drift
ridge (oceanic)

subduction
degassing
Milankovitch theory
eccentricity
precession
obliquity
forcing factor
sulfate aerosols
dimethylsulphide (DMS)
nuclear winter
sunspots
faculae
sunspot cycle
Maunder minimum
conveyor belt (ocean)
quasi-biennial oscillation
radiative equilibrium
radiative forcing agents
desertification

Teaching Suggestions

1.   Climate change is an area of active research at present and receives a lot of coverage on television, in newspapers, and in popular magazines. Use current news or articles to motivate or stimulate discussion in class. *Science News* magazine is a good source of up-to-date information on recent research results. At some universities, it may be possible to bring in a guest speaker that will describe their research and how it relates to climate change.

2.   Examine long-term temperature and precipitation changes at several cities of the students' choice, using the National Climatic Data Center's "CLIMVIS" (Climate Visualization) web site, www.ncdc.noaa.gov/onlineprod/drought/xmgr.html. Do these trends present evidence of global warming? Of global cooling?

Student Projects

1.   Have students obtain climatological data (mean temperature and precipitation amounts), from their city or a city nearby, for as long a period of time as possible. Are there any noticeable periodic variations or long-term trends? Can any observed fluctuations be correlated with recent volcanic eruptions, strong ENSO events, or the sunspot cycle? Where were the weather data obtained? Has urbanization had any effect on the weather data?

It might be possible to obtain a section from a recently cut tree. Then, if a date can be accurately identified with one of the growth rings, try to determine whether the width of the rings is correlated with past mean annual temperatures or yearly rainfall amounts.

2.   Using a book that describes the local geology, have the students prepare a report on the past climate in their region. If possible, photograph or obtain samples from geological structures that were used to infer past conditions.

Blue Skies 3.   Use the Atmospheric Chemistry/Temperature Trends section of the BlueSkies cdrom to examine estimates of future global temperatures. According to these model projections, is the term "Global Warming" appropriate? Which areas of the globe are projected to warm the most? Are there any areas that are expected to cool?

Blue Skies 4.   Using the Weather Forecasting/Forecasting section of the BlueSkies cdrom, examine the relationship between current temperatures and cloud cover. Describe any relationship you observe. Using this observation, comment on the possible effects that future changes in cloud cover might have on future temperature changes.

Answers to Questions for Thought

1.   The chemical concentrations retrieved from the water and air bubbles within ice cores may have undergone a variety post-depositional processes which can complicate their interpretation. These processes include chemical reactions, blowing and drifting, wind pumping (horizontal movement within the snowpack), freezing and thawing, and folding associated with subsurface glacial movement.

2.   Because of the lower temperatures (in the Northern Hemisphere), precipitation was probably less than at present.

3.   The effect of an increase in global cloudiness would probably depend upon the type and height of clouds that form. High, thin cirriform clouds would probably promote global warming (positive feedback) by allowing much of the incident sunlight to pass through them. At the same time, these cold clouds would radiate back to earth more infrared radiation than they would emit to space. Low stratified clouds would probably cool the planet (negative feedback) by reflecting much of the incident sunlight back to space. Also, their warm tops would radiate away much of the infrared energy that they receive from the earth.

4.   For snow to fall, the surface air temperature need only be near freezing. In fact, the heaviest snows generally occur when the air temperature is only slightly below freezing. The main factor in keeping the snow on the ground for an entire year is the air temperature in summer. Cooler summers, of course, produce less melting and a greater likelihood that the snow will exist (in certain areas) for a long time. Once the snow stays on the ground it produces even cooler summers by reflecting much of the sunlight that strikes the region.

5.   Sulfate aerosols may scatter incident sunlight back into space. Sulfate aerosols may also serve as cloud condensation nuclei and thereby modify clouds. With high condensation nuclei concentrations, clouds are composed of a larger number of smaller droplets and are more reflective.

6.   In the Northern Hemisphere ice ages are more likely when the tilt of the earth is at a minimum. During this time, summers would be cooler and less snow would melt. For winter snow to fall, temperatures need only be at freezing or below. In fact, during the slightly warmer winter, more snow

may actually fall (often there is a reduction in snow fall during very cold winters). By the same reasoning, ice ages are more likely when the sun is closest to the Northern Hemisphere during the winter.

7.   Warmer poles would decrease the equator-to-pole temperature gradient, thereby reducing the strength of midlatitude winter storms.

8.   If the oceans remove less CO_2 from the atmosphere, then atmospheric concentrations of CO_2 could be expected to increase. This could cause the earth to warm further.

Answers to Problems and Exercises

1.   0.38 cm of precipitation is 1.9 meters of snow (using a water equivalent of 1 to 5). 3000 meters divided by 1.9 meters/year is about 1580 years.

Multiple Choice Exam Questions

1.   If the average snowfall over an area of north central Canada is 100 cm per year, how long would it take for the snow to reach a depth of 1000 meters? (Assume that there is no melting in summer and no compaction)
a. 10 years
b. 1,000 years
c. 10,000 years
d. 100,000 years
e. 10,000,000 years

ANSWER: b

2.   Ice cores record both the record of past temperatures, and also the causes of climate change.
a. true
b. false

ANSWER: a

3.   The higher the ratio of oxygen 18 to oxygen 16 in the shells of organisms that lived in the sea during the geologic past, the ___ the climate at that time.
a. colder
b. warmer
c. wetter
d. drier

ANSWER: a

4.   Which of the following is <u>not</u> true?
        a. oxygen 16 evaporates more readily from the ocean than oxygen 18
        b. oxygen 16 and oxygen 18 are found in roughly equal amounts in ocean water
        c. the nucleus of oxygen 18 contains two more neutrons than the nucleus of
           oxygen 16
        d. both oxygen 16 and oxygen 18 are found in the shells of marine organisms

ANSWER: b

5.   A high concentration of oxygen 16 found in the ice caves of Antarctica and Greenland would
        indicate ___ at the time the ice was formed.
        a. cold air temperatures
        b. mild winters
        c. intense ultraviolet radiation
        d. the caves were under the ocean

ANSWER: a

6.   Indications of past climates in tree rings are determined by examination of the
        a. thickness of growth rings
        b. density of growth rings
        c. presence of frost rings
        d. all of the above
        e. only a and b

ANSWER: d

7.   Evidence suggests that throughout much of the earth's history, the global climate was:
        a. warmer than it is today
        b. colder than it is today
        c. about the same temperature as it is today
        d. more variable than it is today

ANSWER: a

8.   Which of the following has been used to reconstruct past climates?
        a. analysis of air bubbles trapped in ice
        b. study of documents describing floods, droughts and crop yields
        c. the ratio of oxygen 18 to oxygen 16 in the shells of marine organisms
        d. study of geologic formations
        e. all of the above

ANSWER: e

9. Presently glaciers cover ___ of the earth's surface.
   a. about 10 percent
   b. less than 1 percent
   c. approximately two thirds
   d. none

ANSWER: a

10. During the Pleistocene epoch:
    a. continental glaciers <u>continuously</u> covered large parts of North America and Europe
    b. it was much warmer than now
    c. continental glaciers alternately advanced and retreated over large portions of North America and Europe.
    d. tropical vegetation was growing over vast regions of the Central Plains of North America.

ANSWER: c

11. Thick sheets of ice advanced over North America as far south as New York as recently as:
    a. 1816 ("the year without a summer")
    b. 1550
    c. 18,000 to 22,000 years ago
    d. 2 million years ago, at the beginning of the Pleistocene epoch

ANSWER: c

12. During the climatic optimum:
    a. the climate favored the development of certain plants
    b. average temperatures were colder than at present
    c. continental ice sheets began to melt
    d. alpine glaciers began to advance down river valleys

ANSWER: a

13. Over the past 100 years or so, it appears that average global temperatures have:
    a. increased slightly
    b. fluctuated widely but shown no overall change
    c. decreased slightly
    d. remained constant

ANSWER: a

14. The Viking colony in Greenland perished during:
   a. the Pleistocene epoch
   b. the climatic optimum
   c. the Little Ice Age
   d. the explosion of Mt. Pinatubo

ANSWER: c

15. The Medieval Climatic Optimum was a relatively _____ period.
   a. cold
   b. warm

ANSWER: b

16. During the Little Ice Age:
   a. the climatic optimum occurred
   b. the Bering land bridge formed
   c. alpine glaciers grew in size and advanced
   d. continental glaciers covered large portions of North America
   e. sea level lowered by about 280 ft

ANSWER: c

17. The "year without a summer" (1816) may have been caused by:
   a. soot from coal fires
   b. particulate matter and gases from volcanoes
   c. a dust cloud produced when a meteorite collided with the earth
   d. deforestation

ANSWER: b

18. If the earth were in a cooling trend, which process below would most likely act as a positive feedback mechanism?
   a. increasing the snow cover around the earth
   b. increasing the water vapor content of the air
   c. decreasing the amount of cloud cover around the globe
   d. increasing the carbon dioxide content of the air

ANSWER: a

19. If the earth were in a warming trend, which of the processes below would most likely act as a negative feedback mechanism?
   a. increasing the water vapor content of the air
   b. increasing the snow cover around the earth
   c. decreasing the amount of cloud cover around the globe
   d. increasing the carbon dioxide content of the air

ANSWER: b

20. A positive feedback mechanism:
   a. acts to reinforce an initial change
   b. acts to weaken or oppose an initial change
   c. will cause a positive change
   d. will cause a negative change

ANSWER: b

21. Which theory explains how glacial material can be observed today near sea level at the equator, even though sea level glaciers probably never existed there?
   a. Milankovitch theory
   b. Theory of plate tectonics
   c. Volcanic dust theory
   d. Maunder theory

ANSWER: b

22. Plate tectonics "explains" certain climatic changes by showing that these changes may be related to:
   a. mountain building
   b. the amount of CO_2 and H_2O released into the atmosphere
   c. the paths taken by ocean currents
   d. the position of the continents
   e. all of the above

ANSWER: e

23. The Milankovitch Theory proposes that climatic changes are due to:
   a. variations in the earth's orbit as it travels through space
   b. volcanic eruptions
   c. changing levels of CO_2 in the earth's atmosphere
   d. particles suspended in the earth's atmosphere

ANSWER: a

24. During a period when the earth's orbital tilt is at a minimum, which would probably not be true?
   a. there should be less seasonal variation between summer and winter
   b. more snow would probably fall during the winter in polar regions
   c. there would be a lesser likelihood of glaciers at high latitudes
   d. there would be less seasonal variations at middle latitudes

ANSWER: c

25.   Precession of the equinox refers to:
       a. changes in the shape of the earth's orbit as the earth revolves around the sun
       b. changes in the tilt of the earth as it orbits the sun
       c. changes in the seasons, especially from winter to summer
       d. the wobble of the earth on its axis

ANSWER: d

26.   The Milankovitch cycles in association with other natural factors explain how glaciers may advance and retreat over periods of:
       a. hundreds of millions of years
       b. several million years
       c. hundreds of thousands of years
       d. ten thousand years to one hundred thousand years
       e. hundreds of years

ANSWER: d

27.   The formation of continental glaciers over vast areas of North America is most favorable when Northern Hemisphere summers are __ and winters are __.
       a. cool, extremely cold
       b. cool, mild
       c. warm, extremely cold
       d. warm, mild

ANSWER: b

28.   Volcanoes that have the most impact on global climate seem to be those rich in:
       a. nitrogen
       b. water vapor
       c. carbon dioxide
       d. sulfur
       e. oxygen

ANSWER: d

29.   Large volcanic eruptions with an ash veil that enters the stratosphere, tend to __ at the surface.
       a. increase temperatures
       b. increase precipitation
       c. decrease temperatures
       d. have no effect

ANSWER: c

30.   The quasi-biennial oscillation is a pattern of:
        a. ice advances and retreats spanning 2 million years
        b. 22 year-long sunspot cycles
        c. reversing stratospheric winds above the tropics
        d. increases and decreases in carbon dioxide concentrations
        e. solar activity that takes about 2 years to complete its cycle

ANSWER: c

31.   It now appears that if global temperatures continue to rise, the oceans:
        a. will increase the amount of CO_2 in the atmosphere
        b. will reduce the amount of CO_2 in the atmosphere
        c. could increase or reduce the amount of CO_2 in the atmosphere
        d. will have no effect on atmospheric CO_2 concentrations

ANSWER: c

32.   It appears that, as the number of sunspots increases, the sun's total energy output:
        a. increases slightly
        b. decreases slightly
        c. begins to alternately increase and decrease above average values
        d. does not change

ANSWER: a

33.   The Maunder Minimum refers to a time when:
        a. the earth was in the middle of an ice age
        b. the tilt of the earth's axis was less than it is now
        c. the earth was closer to the sun than it is now
        d. few snowstorms occurred over the United States
        e. there were fewer sunspots on the sun

ANSWER: e

34.   Studies reveal that during colder glacial periods, CO_2 levels _____ during warmer interglacial periods.
        a. were higher than
        b. were lower than
        c. were about the same as
        d. were more variable than

ANSWER: b

35. For CO_2 to produce a global warming of between 2 °C and 5 °C, climatic models predict that the concentration of this gas <u>must</u> also increase in the atmosphere.
    a. chlorofluorocarbons (CFCs)
    b. nitrous oxide (N_2O)
    c. water vapor (H_2O)
    d. ozone (O_3)

ANSWER: c

36. An increase in atmospheric CO_2 concentrations will most likely lead to ___ in the troposphere and ___ in the upper atmosphere.
    a. warming, warming
    b. cooling, warming
    c. warming, cooling
    d. cooling, cooling

ANSWER: c

37. It is now known that, overall, clouds:
    a. have a net warming effect on climate
    b. have a net cooling effect on climate
    c. have no net effect on climate
    d. are the single most important feature in determining climate

ANSWER: b

38. The most recent warming trend experienced over the Northern Hemisphere could be the result of:
    a. increasing volcanic eruptions
    b. light colored particles in the stratosphere
    c. increasing levels of CO_2
    d. a decrease in the energy emitted by the sun
    e. an observed decrease in snow cover

ANSWER: c

39. Everything else being equal, a gradual increase in global CO_2 would most likely bring about:
    a. an increase in surface air temperature
    b. a marked decrease in plant growth
    c. a decrease in evaporation from the earth's oceans
    d. no change in global climate

ANSWER: a

40. Which of the conditions below would <u>most likely</u> produce warming at the earth's surface?
   a. increase the amount of low-level global cloudiness
   b. increase the amount of sulfur-rich particles in the stratosphere
   c. decrease the energy output of the sun
   d. increase the amount of global snow cover
   e. increase the amount of high-level global cloud cover

ANSWER: e

41. Which below is <u>not</u> one of the possible consequences of global warming predicted by climate models?
   a. accumulations of additional snow in Antarctica
   b. a reduction in average precipitation over certain areas
   c. lowering of sea levels
   d. a cooling of the upper atmosphere
   e. a drop in the rate of ozone destruction in the stratosphere

ANSWER: c

42. The much-studied temperature record of the past 140 years is derived from
   a. land-based observations only
   b. oceanic observations only
   c. both land-based and oceanic observations

ANSWER: c

43. A biogeophysical feedback mechanism in the Sahel relates:
   a. reduced cloud cover to increasing surface temperatures and an increase in rainfall
   b. reduced vegetation to a decrease in surface temperatures and a reduction in rainfall
   c. an increase in vegetation to an increase in surface temperatures and a reduction in rainfall
   d. a decrease in vegetation to a lowering of surface albedo and an increase in rainfall

ANSWER: b

44. A significant percentage of the observations comprising the much-studied temperature record of the past 140 years were not air temperatures, but rather were temperatures of ocean surface waters.
   a. true
   b. false

ANSWER: a

344

45. The ocean's conveyor belt circulation is controlled by
   a. gradients in temperature
   b. gradients in salinity
   c. gradients in density
   d. all of the above

ANSWER: d

Essay Exam Questions

1. Describe some of the techniques used to infer past climates. About how far into the past are these different methods able to reconstruct past climatic conditions?

2. Explain how analysis of glacial and polar ice cores provide important clues about past climate.

3. Briefly describe the climatic variations that have occurred since the beginning of the Pleistocene epoch to the present day.

4. Asia and the North American continent have been connected by a land bridge at various times during the past. When was the Bering land bridge most recently exposed? Did this coincide with a warm or cool climatic period?

5. When did the Little Ice Age occur? Have there been any suggestions of a possible cause of the Little Ice Age?

6. What are meant by the terms positive and negative feedback mechanisms? Give an example of a process that would be considered a positive feedback mechanism during a period of warming on the earth. Can you think of a negative feedback mechanism?

7. Describe one positive and one negative feedback mechanism associated with human health or the human body.

8. Discuss the changes that occur in the earth's orbit and orientation of the earth in its orbit. Over what time scales do these different changes occur?

9. What combination of the earth's orbital eccentricity, and precession and obliquity of the earth's axis would produce hottest summer temperatures in the Northern Hemisphere?

10. In what ways might global warming benefit the planet?

11. Assume that you have 100 years of continuous temperature records from you local weather service office. Discuss some of the difficulties you might have trying to determine whether average temperatures have increased during this period.

12. Carbon dioxide concentrations in the atmosphere have been increasing steadily since the beginning of the industrial revolution and may double by the middle of the next century. Why is this of concern?

13. Does deforestation act to increase or decrease atmospheric carbon dioxide concentrations?

14. In addition to carbon dioxide, what other gases may contribute to the problem of global warming? Why might global warming lead to an increase in global cloudiness? What possible effects would increased cloudiness have on global warming?

15. What are some of the possible consequences of global warming?

16. Where is the Sahel located? Is this a region of scarce, variable or abundant rainfall? Describe the biogeophysical feedback mechanism whereby a reduction in surface vegetation might cause a reduction in precipitation amounts.

17. Discuss the significance of a predicted increase in global temperatures of 3 °C, as opposed to a predicted increase of 1 °C.